W0115995

AN INTRODUCTION TO THE LOGIC OF THE COMPUTING SCIENCES

A Contemporary Look at Symbolic Logic

Richard F. Von Dohlen

University Press of America,® Inc.
Lanham • New York • Oxford

Copyright © 1999 by
University Press of America,® Inc.
4720 Boston Way
Lanham, Maryland 20706

12 Hid's Copse Rd.
Cumnor Hill, Oxford OX2 9JJ

All rights reserved
Printed in the United States of America
British Library Cataloging in Publication Information Available

Library of Congress Cataloging-in-Publication Data

Von Dohlen, Richard F.
An introduction to the logic of the computing sciences : a
contemporary look at symbolic logic / Richard F. Von Dohlen.
p.   cm.
1. Computer logic. 2. Computer science. I. Title.
QA76.9.L63V66 1999   004—dc21   98-51878 CIP

ISBN 0-7618-1326-8 (pbk: alk. ppr.)

♾™ The paper used in this publication meets the minimum
requirements of American National Standard for Information
Sciences—Permanence of Paper for Printed Library Materials,
ANSI Z39.48—1984

Contents

Preface

An Introduction to the Logic of the Computing Sciences: A Contemporary Look at Symbolic Logic grew out of my teaching students with diverse academic and professional interests: philosophy, mathematics, computing sciences, pre-law, engineering, the social sciences, etc. I needed a text that would make connections between symbolic logic and all of these diverse fields. I have attempted this by the use of two basic strategies. Each principle is introduced by explanations in ordinary English, truth tables, flow charts and algorithms in order to show the relationship between symbolic logic and the computing sciences. Secondly, I have included numerous word problems on such diverse topics as legal liability, medical treatment, artificial intelligence, figuring of pay scales, the making of sound business judgments, the nature of love, the goodness and omnipotence of God, the social theories of Marx and Spencer, and the belief systems of different Muslim groups. Life is more than logic and logic is more than symbolic logic. Nevertheless, the structure under girding symbolic logic is interwoven with life in interesting and enriching ways. The computing sciences have become a part of all of our lives whatever our professions or belief systems. Understanding symbolic logic can help us understand something about how human beings think. Furthermore, understanding symbolic logic can enable us to understand much about "how computers think." This is what I have sought to demonstrate. Thus, this is an introductory text that emphasizes connections.

This text differs from many in at least one other important respect. Many good students find symbolic logic foreign and intimidating. For these students it is important not just to convince them of the importance of logic but also to persuade them that it is a subject that they can master. Texts that introduce eight or nine principles at a time followed by symbolic and word problems that require use of all of these principles discourage such students. Therefore, in almost every chapter, I introduce one principle at a time followed by ten symbolic problems and ten word problems. The student is asked to do twenty problems that require the use of *modus ponens* and then in the next chapter is asked to complete twenty problems that require the use of both *modus ponens* and *modus tollens*, etc. A study guide containing solutions to 30% of the problems in the text enables students to assess for themselves whether they have understood the process up to that point. Instructors who desire solutions to all of the problems in the text may obtain this for a nominal fee on hard copy and diskette by writing to the author at Lenoir-Rhyne College, Hickory, North Carolina 28603. Please identify your position and make your request on departmental stationary.

Students learn differently. For some the explanation of a principle in ordinary English is most effective. For others, a diagram best enhances the process of understanding. I introduced flow charts, algorithms and truth tables to show the connection between symbolic logic and the computing sciences. A by-product of this approach is enhanced pedagogical effectiveness. Coupled with a gradual introduction of principles and repeated opportunities to practice the application of the principles, this text should be pedagogically sound for all students regardless of their academic major or professional aspirations.

I wish to thank my many logic students over the years for their help and encouragement. I have written this text for them. My wife Nancy has had a distinguished career as a K-through-second grade teacher. She combines loving sensitivity and a commitment to the imparting of basic skills with the conviction that children can and must learn critical thinking skills in their early years. Her job is more important, and she is a better teacher of logic than I will ever be. She often practices her critical thinking skills on me. When asked where a particular article of clothing can be found, she frequently replies not by giving a direct answer but by giving a set of premises from which a thoughtful husband can deduce the answer. Logic, thus, is not just for computers. It can enhance one's marriage as well. The writing of this text would not have been possible without the competent and patient help of Mrs. Saundra Cooke who for many years was secretary to the department of Religion and Philosophy. Without "Sonnie" I would not have attempted the task. Without her hundreds of hours of hard and able work it would not have been completed.

I would also like to thank Lenoir-Rhyne College and the Division of Higher Education and Schools of the Evangelical Lutheran Church in America for their support in this project.

CHAPTER 1: INTRODUCTION

I. What is Logic?

A. Logic and the Disciplines

Logic is the study of correct reasoning. All of us have been learning to reason at least since the time we began to learn to speak. Much of what a student studies in college is designed not so much to learn new "facts" but to help him or her learn to reason effectively in particular situations or with respect to a particular subject matter. Reasoning correctly has always been important. Since, the invention of the printing press, however, it has become increasingly less important to use the mind to "store facts" and more important to discern the relevant from the irrelevant, the true from the false, the good argument from the bad.

The invention and widespread use of the computer has made this more the case than ever. The problem is not to store information but to process it. Learning has less and less to do with memorization of facts and more and more to do with effective and creative use of information.   Effective and creative use of information requires something more than correct reasoning. Analogously, a truly great pianist must have something more than just the ability to correctly read a musical score and hit the right notes. Even great musicians, however, though they must have something more, still must be able to correctly read that musical score and correctly play those notes. Effective and creative use of information may require something more than correct reasoning. Correct reasoning is, none the less, essential for effective thinking.

Every discipline may be said to have its own terminology, rules, methodology, and assumptions, in one sense of the term - its own logic. Thus, chemists, biologists, historians, sociologists and economists must learn the logic of these special disciplines in order to think effectively and creatively in these particular areas. Nevertheless, despite the different ways of thinking required in understanding mathematics, and let us say poetry, computer science and music theory, physics and art criticism, business and theology, psychology and journalism; all of these may be said to some degree to have a logic in common. It is this common logic which will be the object of our study.

B. Different Types of Logic

Many typical texts in logic will be divided into several sections: (1) Logic and Language; (2) Traditional or Aristotelian Logic; (3) Symbolic Logic; and (4) Inductive Logic. Unfortunately, this is more material than can comfortably be covered in a one semester course. This text will concentrate on Symbolic Logic. I will merely attempt to say enough here to indicate in a superficial way what is involved in the other areas.

1. Logic and Language

Logic and language will typically treat problems of the nature of language, different ways of defining terms, various types of meaning, etc. It will also have a section on what is known as informal fallacies or types of fallacious reasoning which we encounter in everyday dialogue; "You should vote for his innocence because he is poor" (appeal to pity); "Since everybody is doing it, it must be okay" (appeal to mass opinion); "If you don't agree with me, I'm going to excommunicate you as a heretic" (argument from force); etc.

Study of this area of logic can be extremely valuable. It has immediate practical consequences and some schools devote an entire course to such a study. You have probably been exposed to some of this material in social science courses and other courses which analyze arguments associated with controversial issues.

2. Deductive and Inductive Logic

Logicians distinguish between deductive arguments and inductive arguments. In brief, a deductive argument is one in which, if the premises are true then the conclusion <u>necessarily</u> follows. For example:

(1) All people exposed to the measles catch them.
(2) Jones has been exposed to the measles.
(3) Therefore, Jones has caught the measles.

If the premises (1) and (2) **were** true, then the conclusion (3) would necessarily be true. Thus, the above is an example of a valid deductive argument. It may not be what logicians call a sound argument. Either premise (1) or (2) or both may be false. None the less, **if** (1) and (2) were true (3) must follow.

An inductive argument is one in which the premises (if true) make the conclusion not necessarily true but only more probable. For example:

(1) Most people exposed to the measles catch them.
(2) Jones has been exposed to the measles.
(3) Therefore, Jones probably has the measles.

Premises (1) and (2) (if true) make the conclusion (3) more probable but not certain. It would be possible for (1) and (2) to be true and (3) to be false without there being anything wrong with the argument.

A great many of the assertions we take as "facts" are conclusions of one or more inductive arguments. Consider the following statements: "Julius Caesar was assassinated"; "Nixon lied about Watergate"; "Smoking causes lung cancer"; "Challenger was destroyed because of defective O-rings"; and "The sun will rise tomorrow." All of the statements given above are held to be true by most people. If we were to examine the arguments given in support of these assertions we would frequently (if not in every case), find (1) statements of fact and

(2) assertions that since the facts being asserted are true, it is **probably** the case that some other statement is true. For example: Since we have discovered that there is a statistical correlation between smoking and lung cancer we reason that smoking is a probable cause of lung cancer. For most of us the evidence is so conclusive that we don't doubt the relationship. The American Tobacco Institute, however, protests that a conclusive link (do they perhaps mean a deductive relationship?) between smoking and lung cancer has not been shown. Furthermore, since smoking makes lung cancer only more probable, it is not necessarily the case that if you smoke you will get lung cancer. Hence, millions continue to smoke.

The study of the structure of deductive arguments is called Deductive Logic. The study of inductive arguments is called Inductive Logic. Inductive Logic is extremely important and it is perhaps regrettable that we cannot treat it this text. Inductive Logic, however, like the discussion of logic and language is treated in other courses required or elective within most college curriculums. Thus, although, you may not engage in a formal study of Inductive Logic, you will be exposed to it in other contexts.

3. Types of Deductive Logic

Logicians typically divide discussion of deductive logic into "classical" or "Aristotelian" logic (so designated because it was first developed and systematized by the philosopher Aristotle) and "modern" or "modern symbolic" logic. Aristotelian logic is a remarkable achievement and certainly worth study. Furthermore, it is still widely used and is probably the logic to which people educated in the fifties and sixties were exposed.

Modern or symbolic logic was developed around the turn of the century by a number of thinkers. Alfred North Whitehead and Bertrand Russell are the figures most prominently associated with it. I am least concerned with the neglect of Aristotelian logic in favor of symbolic since in my view symbolic logic is both far more useful and powerful than Aristotelian. Since the whole course in a sense will be an explanation of symbolic logic I will delay a description of its basic approach until we begin our exposition of the fundamentals of a logical system.

C. Philosophy of Logic

Many of the issues related to the study of logic are controversial, at least among philosophers. For example, the relationship of logic to reality, the relationship of logic to language or the relationship of logic to a particular discipline like mathematics. For example, Russell and Whitehead made several assumptions and set out to prove a number of assertions. They believed that the best way to get at reality was not through the use of ordinary language which tends to be vague, ambiguous, lacking in specificity and unclear in various other ways. Rather, what was needed was an artificial language which could be precise and clear and lend itself to rigorous deductive systematization. This was one of the main motives behind the development of modern logic.

The relationship of ordinary language to reality is an interesting problem which we regrettably will not have much time to discuss here. I am going to "assume" for purposes of this text that although "life is more than logic" and certainly more than symbolic logic, that the systematic study of logic will be useful in helping one to better understand the reality in which we "live and move and have our being." Such an assumption is necessary if we are to proceed with this text. We will not, however, take the time to debate the philosophical issues surrounding that assumption.

II. What is the Payoff?

Who should take logic? What may one expect to get out of it? Is it hard? Is it fun? Is it valuable? If I'm not good in mathematics will I have a rough time in logic? Will it be worth my time? The best way for a student to get answers to these and other similar questions is not to ask the professor but to ask one or more students who have already had a course in symbolic logic. It is not that college professors are in general dishonest but they certainly tend to be biased in favor of what ever it is that they are teaching. But asking other students is not foolproof either. Students differ in their interests, goals, and abilities. Your friends' favorite course may turn out to be one you hate and vice a versa. Despite these difficulties let me attempt to give you some direction and information.

Logic is the study of correct reasoning. Therefore, anyone interested in reasoning more effectively can profit from the study of logic. This is not to say that only those who have had a course in logic always reason correctly. Nor is it the case that the study of logic will necessarily make you a brilliant debater. This is comparable to the study of other disciplines. A course or even a Ph.D. in psychology will not necessarily make you emotionally healthy. A biology course does not insure physical health. All business majors are not assured of becoming rich. A physical education major does not automatically become a world class athlete. Conversely, there are emotionally and physically healthy people, successful business men and women and world class athletes who have never engaged in formal study of the respective areas cited above. In all endeavors there is no substitute for natural ability, good fortune and informal experiential learning.

My own experience, however, and that of many students I have taught over the years is that a formal study of logic does significantly enhance one's reasoning powers. By the time you have mastered the material in this text, you should be able to reason more effectively in a wide variety of areas and analyze with ease complex arguments which would have totally confused you at the beginning of your study.

The study of formal logic is a requirement for anyone doing serious work in philosophy. It is very useful for anyone studying computer science. It is almost universally recommended for anyone who plans to go to law school and to take the Law School Admissions Test (LSAT). Anyone planning to do serious theological study should definitely have some background in formal logic.

Is it hard? Some people find it hard. Many find it very easy. It is important if you are taking a one semester course using this text that you do all the homework and keep up. Cramming at the last minute is not recommended for any discipline. It definitely is counter productive in logic. If you keep up, however, you will probably be surprised at how easy it is.

What if I'm not good in math? It doesn't matter. It is true that the study of mathematics requires rigorous deductive reasoning. Therefore, people who are good at mathematics will frequently do well in logic. The reverse, however, is not necessarily true. The text requires no background in the formal study of either logic or mathematics. Furthermore, although we will be concerned with symbolization and deduction we will be working a lot with ordinary English sentences.

Some of my very best logic students have had a poor background in mathematics.

Will it be worth my time? I have designed this text toward that end. By the time you have mastered the material in this text you should be able to reason more effectively. You should have a better command of the English language and be able to express yourself more clearly. You will be able to take apart complex arguments which heretofore would have intimidated you. Those of you who are taking courses in computer programming should find your understanding and facility at programming enhanced. Those planning to take the LSAT should find this text excellent preparation.

III.   The Design and Purpose of the Text.

Those taking Introductory logic tend to be divided into several groups. Majors in philosophy, theology/philosophy, mathematics, and computer science take logic because of its direct relevance to their specific majors. Students planning to go to law school (whatever their majors) take logic in order to improve their chances of doing well on the LSAT. Others take logic simply because they are interested in improving their reasoning powers. I know of no published text designed to meet the needs of a general audience but also specifically designed to meet the needs of all these groups. This text will attempt to do just that.

In this text we will concentrate on symbolic logic. Many logic texts try to cover all aspects of the field. This can have a number of unfortunate results. Students, buy more text than they use. Beyond that, however, material is sometimes introduced in large chunks. This is fine for the very best students. Potentially good students, however, find it indigestible and are often easily discouraged. I endeavor in this text to introduce material gradually and in digestible portions. Thus, if a student is having trouble, student and instructor will be better able to pinpoint exactly at what point the difficulty begins.

This text will contain the minimum of technical terminology necessary for understanding the material. Some texts even though they are introductory in nature read as though they were written by specialists in logic for other specialists. This text will assume that most of you want to use logic to understand other phenomena and problems and not to become specialists in the field. At the same time, it should serve as a good introduction to those who wish to pursue further the field of logic as a specialized study.

This text has several unique features as compared to others which might have been chosen by your instructor. Perhaps the most unique feature of the text is its explicit attempt to relate symbolic logic to techniques and structures applicable to computer programming. To this end, each new principle is introduced by explanations of four different types: ordinary English, truth tables, flow charts and algorithms.

Secondly, the text follows the plan of introducing new concepts in small bites, generally one new concept at a time. Each new section contains a set of twenty problems. This includes a set of ten problems which are already in symbolic form and ten word problems, which must first be translated into symbolic form before they can be solved. The difficulty of the problems ranges from the most simple to very complex. By the time the student has completed these twenty problems the new concept will be fixed firmly in his or her mind. Thirdly, the word problems deal with a variety of topics dealing with business payrolls, theological disputes, ethical issues and legal relationships. These word problems have been designed to enhance the students ability to deal with the sort of problems prominent in the admissions tests in law, business, medicine such as LSAT, GMAT, MCAT, etc. Thus, study of this text will be an excellent supplement to those books designed to prepare students to take these tests but it should not be regarded as a substitute for such guides.

Finally, symbolic logic is a sub-discipline within the field of philosophy. It also provides tools of analysis which are used in all of the other areas in philosophy. The author of this text has attempted to suggest the relationship of logic to the other areas of philosophy by introducing word problems related to some of these other areas i.e., philosophy of religion, ethics, philosophy of the social sciences, epistemology, metaphysics, etc.

CHAPTER 2: LOGIC, PROBLEM SOLVING and ALGORITHMIC THINKING: Some General Comments on How to Think Logically

The techniques learned in this course will not provide an exhaustive account of how to learn to think. Such an account is the appropriate domain of the field of psychology of learning. Logic is a sub-discipline of philosophy rather than the field of psychology. As such, it is concerned with the study of the science of correct reasoning not in the study of **how** we come to learn to reason. Nevertheless, logic is studied by students planning to go into such fields as law, medicine, engineering, computer science, journalism and business, as well as theology and philosophy because it is presumed that an introductory course in logic will help one to learn to think and solve problems in these areas. I think this assumption is well founded. Learning the techniques of problem solving necessary for the successful mastery of the material in this course can assist in the mastery of problems in a variety of other areas.

To be successful in logic, memorization of certain rules and mastery of certain techniques is critical. At least as important, however, is the adoption of a **way of thinking** which is already quite natural for some of my readers but which **can become natural** for others with a little effort. We will refer to this way of thinking as "algorithmic thinking." This is a method of problem solving which sets out a detailed plan listing a finite number of steps leading to the solution of a problem.

Ideally, a complete algorithm would give an exhaustive set of directions for solving a problem which can be followed in a purely mechanical fashion. Computer programs require such exhaustive sets of directions. Legal documents like wills, insurance policies, real estate deeds, etc. aspire to such completeness. We pay attorneys to create documents which give clear and exhaustive directions which can be followed in a purely mechanical fashion. We also pay them to contest the legality of documents which fail to achieve this ideal.

A completed solution to a logic problem will be for that problem a complete algorithm in the sense described. When we speak of "thinking algorithmically" we are attempting to encourage the kind of thinking which leads to the creation of algorithms. This text will throughout suggest strategies and introduce problems which will enhance your ability to create algorithms. The assumption is that the ability to create algorithms is either helpful or absolutely necessary in dealing with more complex legal and philosophical problems and in creating complex computer programs. Thinking algorithmically is extremely important but can, I believe, be illustrated quite simply.

When the author of this text was ten years old, he lived one year in the village of Spruce Head, Maine, with a population of about four hundred people. Assuming we were both in that village, an acceptable set of directions for finding the house I lived in would be as follows: (1) Get in your car; (2) Start your engine and drive around until you find a yellow house on the main road overlooking a salt water cove.

Better directions are possible but these would be adequate. But now let us suppose that we set for ourselves a different task. Let us assume that we want to get from Lenoir-Rhyne College in Hickory, NC. to the University of Chicago. Even supposing a very clear and complete description of the University of Chicago, instructing one to get into his or her car and drive around until finding the University of Chicago is obviously not adequate. We need something a bit more complicated.

We would want to plan our journey and to make or have made for us a plan or map directing us from Lenoir-Rhyne to the University of Chicago. Assuming we had a fairly detailed map of the United States, and perhaps a series of detailed maps of states, cities and towns within the United States, our plan would go something like this.

1. Discover that Lenoir-Rhyne is in Hickory, North Carolina.
2. Discover that the University of Chicago is in Chicago, Illinois.
3. Find Illinois on the map.
4. Find Chicago in Illinois.
5. Find North Carolina on the map.
6. Find Hickory in North Carolina.
7. Notice that Chicago is North and West of Hickory.
8. Trace highways from Hickory to Chicago.

This assumes (a) that we are driving and not flying, (b) that we want the most direct driving route to Chicago.

This procedure will fairly quickly lead to something like the following observations: If we can get on route #94 in Indiana heading west it will take us to Chicago. Getting on route #65 N at Indianapolis will take us to route #94. Route #74 will take us from Cincinnati to Indianapolis. The best way to get to route #74 at Cincinnati is to take route #75 from Knoxville to Cincinnati. Route #40 W will take us from Hickory to Knoxville.

Actually, once we have established our most general plan, we need to break our problem down into smaller problems: (1) Getting from Lenoir-Rhyne College to I-40 heading West; (2) Getting through or around Knoxville and on I-75 heading North; (3) Getting through or around Cincinnati and on I-74 heading North; (4) Getting through or around Indianapolis and on I-65 heading North; (5) Getting from I-65 to I-94; and (6) from I-94 to the University of Chicago. Once we have arrived at the University of Chicago we may have a further goal like finding my friend John Bell's apartment.

Supposing that we were taking a van load of students to the University of Chicago. We might assign each of these problems to different members of our party. Paul might be assigned problem #(1) getting from Lenoir-Rhyne to I-40 heading West. Mary could be assigned problem #(2) getting through or around Knoxville and onto I-75 N. Jeremy problem #(3) and Tom problem #(4), etc. It is conceivable that no one person would have the competence to direct us all the way to John's apartment at the University of Chicago but that collectively we could get along just fine. This provides us with an analogy to some large computer programs with many sub-programs worked on by a number of different programmers. All of this would be possible only if at least one person in the group had general knowledge of the overall plan and instructed the others accordingly.

We will be developing numerous strategies for problem solving as we go along. How shall we go about solving problems in this text? Several possibilities emerge. (1) You could wait until the instructor or another student in the class solves the problem, memorize the steps involved and hope that the problems you have memorized will appear on the exam. There are several reasons for not taking this approach. In the first place, it won't work. There is too much to memorize of too complicated a nature. (2) You could perform random logical operations on the premises until you arrived at the solution. This would probably work on some of the earlier problems where we are dealing with only a few principles. This is not a good plan in the long run. In the first place, as time goes on, we are going to learn more principles and deal with problems with numerous premises. As we do, the number of possible logical operations will rise exponentially. This plan would be analogous to trying to get from Hickory to Chicago by randomly exploring every available road. Given enough time and resources, it would work. But it would be very inefficient. (3) A third strategy is intuition. Review the premises and the logical principles available to you until the answer comes to you. This also works for some people, especially if one is very bright and/or the problem is not too complicated.

As a matter of fact, successful solutions to problems in logic, computer science, engineering, mathematics, law, theology and politics are attributed to elements of memorization, random searching and intuition. These elements cannot be dispensed with nor should they be despised. Memory, random searching and intuition are helped, however, if we can develop a strategy to "get us in the neighborhood," so to speak, of our solution. It is toward the end of developing such a strategy that this text is dedicated.

Four rules readily emerge. (1) Become familiar with the techniques and procedures appropriate to problem solving in the area under consideration. (2) Sketch out the problem in its most general features. Don't start with particular features of the problem. (3) Break the problem down into sub-problems and if necessary sub-sub-problems or if necessary even sub-sub-sub-problems. (4) Start from the conclusion and work backwards, step-by-step, breaking the larger problem down into several smaller problems. Proceed from your premises, step-by-step, to the conclusion. The analogy is not perfect but this is not unlike our illustration of planning our trip from Hickory to Chicago. We start at Chicago and work backwards conceptually until we connect Chicago and Hickory on the map. We then proceed from Hickory to our destination.

CHAPTER 3: STRUCTURE OF ARGUMENTS

Every deductive argument has a structure.

If Jones was negligent, then he will be liable for damages. Jones was negligent. Therefore, Jones will be liable for damages.

If Smith worked more than 40 hours, then he receives overtime pay. Smith worked more than 40 hours. Therefore, Smith receives overtime pay.

If God is non-temporal, then his being is incomprehensible. God is non-temporal. Therefore, God's being is incomprehensible.

If Jim is late for dinner on his anniversary he is going to have an unromantic evening. He is already running an hour late. Jim is going to have an unromantic evening.

The four arguments given above speak of different things--Jones' legal problems, Smith's paycheck, the being of God, and Jim's long and rather frustrating evening. What they have in common is the structure of the argument.

The structure is easily discernible. We can demonstrate this by assigning a capital letter to each portion of a sentence concerning which a relationship is asserted. For example: Let N equal "Jones is negligent." Let L equal "Jones is liable for damages."

If we use an arrow (→) to signify (if …. then), we may symbolize "if Jones is negligent then Jones is liable for damages," as N → L.

Our argument is symbolized in figure 3.1as follows:

Figure 3.1

| N → L |
| N |
| ∴   L |

We may treat the other arguments in a similar fashion. Let H equal "Smith worked more that forty hours." Let O equal "Smith receives overtime pay." Our argument is symbolized in figure 3.2 as follows:

Figure 3.2

| H → O |
| H |
| ∴   O |

Let G equal "God is non-temporal." Let I equal "God's being is incomprehensible." The argument is then symbolized in figure 3.3 as follows:

Figure 3.3

| G → I |
| G |
| ∴   I |

Let J equal "Jim is late for dinner on his anniversary." Let U equal "Jim will have an unromantic anniversary." The argument is then symbolized in figure 3.4 as follows.

Figure 3.4

$$
\begin{array}{c}
J \to U \\
J \\
\hline
\therefore \qquad U
\end{array}
$$

It becomes obvious that these arguments which talk about very different things have the same structure. By symbolizing them in this fashion we are able to focus on the structure of the argument without being distracted by the specific subject matter under consideration. This is important when we are attempting in a cool and rational manner to consider such momentous phenomena as the incomprehensibility of God and an unromantic wedding anniversary.

Logicians have identified a number of different argument structures many of which we will use in this course. They have given these specific names and developed a notation for symbolizing them. In our examples given above we used specific capital letters to symbolize specific statements. We will use lower case letters to signify no particular statements but the structure of the argument itself. Thus, we symbolize the structure of the arguments given above in figure 3.5 as follows:

Figure 3.5

$$
\begin{array}{c}
p \to q \\
p \\
\hline
\therefore \qquad q
\end{array}
$$

This argument form is identified as modus ponens (to place or put) and is abbreviated M.P. .Modus ponens or M.P. is an example of a valid argument form. Later on in our discussion we will consider a method for proving that an argument form is valid. For the time being I will ask you to accept it as valid either because it is intuitively obvious to you that it is valid or failing that because you trust the author of the text not to mislead you.

CHAPTER 4: ORDINARY ENGLISH AND "IF-THEN" STATEMENTS*

The form if p then q (p → q) may be written in English in several different ways. Translating these into symbolic form is sometimes difficult but a clear understanding of the meaning of p → q will be helpful.

1. "If p then q" states a necessary (**but only hypothetical**) relationship between p and q. For example, if I am a man then I am a mammal is a true statement. It happens to be the case that I am both a man and mammal. If I were Rockefeller, then I would be rich is also a true statement even though I happen to be neither rich nor Rockefeller.

2. "If p then q" states a **necessary** relationship between p and q. If p then q means that if we have p then q necessarily, inescapably, inevitably, without fail, etc., follows. Thus, we may speak of **q being a necessary condition of p**.

---

* When you begin to attempt solutions to word problems a little later in the text you will want to return to this section for guidance.

3. If q is a necessary condition of p then p is a **sufficient condition** of q. In other words, whenever we have p, that is sufficient or enough to guarantee that we will have q.

4. If p is sufficient to guarantee q, then it is also true that if p then we **always** will have q. It is also the case that we can have p **only if** we have q.

5. The order in which the p statement and q statement appear in the sentence do not effect the **logical** order. For example: "If p then q" and "q if p", both mean p → q.

6. If p is **sufficient** to guarantee q, and if q necessarily follows from p, then it follows that you cannot have a p **unless** you have a q.

Thus, if we let L equal jumping in the lake and W equal getting wet, L → W might be written in English in a number of different ways.

If you jump in the lake, then you will get wet. You can jump in the lake only if you get wet. You will get wet if you jump in the lake. Jumping in the lake is a sufficient condition for getting wet. Getting wet is a necessary condition for jumping in the lake. Jumping in the lake entails getting wet. Jumping in the lake implies getting wet.

7. In translating ordinary English statements into symbolic form it is critical to recognize that portion of the statement which **functions** as the p statement and distinguish it from that portion of the statement which **functions** as the q statement. For example, take the statement, "If John and Mary go to Spain and France, then they will go to Madrid and Paris by train." The p statement is, "If John and Mary go to Spain and France." The q statement is, "they will go to Madrid and Paris by train."

8.   Sometimes we may have one or more "if-then" statements "nesting" in an "if-then" statement. For example, "If it is true that if John goes to Spain then he will go to Madrid, then he will have a good time."

Let S equal "John goes to Spain." Let M equal "John will go to Madrid." Let G equal "John will have a good time." The statement would be symbolized $(S \rightarrow M) \rightarrow G$

Or even more complicated still: "If being morally responsible means one is legally liable, then if Jones obeys the law then he or she will be treated with greater respect." Let M equal "being morally responsible." "Let L equal "one is legally liable." Let O equal "Jones obeys the law." Let R equal "Jones will be treated with greater respect." This statement would be symbolized as $(M \rightarrow L) \rightarrow (O \rightarrow R)$

9. Depending on the focus of our attention **different** portions of a statement may function as the p or q statement. For example, consider the following argument.

"If being morally responsible entails one is legally liable, then if Jones obeys the law he will be treated with greater respect. Being morally responsible is a sufficient condition for being legally liable. Jones does obey the law. Therefore, Jones will be treated with greater respect." M, L, O, R.

The argument will then be symbolized as below in figure 4.1.

Figure 4.1

	P	Q
1	$(M \rightarrow L) \rightarrow (O \rightarrow R)$	$\therefore$ R
2	$(M \rightarrow L)$	
3	O	
4	$(O \rightarrow R)$	1,2 M.P.
5	R	4,3 M.P.

In step number 1 of figure 4.1, $(M \rightarrow L)$ functions as the p statement and $(O \rightarrow R)$ functions as the q statement. Thus, we can prove $O \rightarrow R$ in step #4 from Proposition number 1 and proposition number 2 and Modus Ponens.

Figure 4.2

		P	$\to$	q	
1		$(M \to L)$	$\to$	$(O \to R)$	
2		$M \to L$			
4	$\therefore$	$O \to R$			

In step number 4 as depicted in figure 4.2, O functions as the p statement and R functions as the q statement. Thus, we can prove R, as shown in figure 4.3, from proposition number 4 (which we have deduced from #1 and #2, with the help of Modus Ponens) and proposition number 3 which was given to us in the original statement of the argument.

Figure 4.3

4		O	$\to$	R
3		O		
5	$\therefore$	R		

10. Ordinary English is often vague, ambiguous, lacking in specificity, and unclear in various other ways. This means that it may not always be clear that an "if-then" statement is being asserted. Thus, careful analysis and interpretation of the English statement is often required. Frequently in our day-to-day conversation this doesn't matter. In legal documents, computer programs, scientific theories and technical arguments in philosophy and theology, however, correct interpretation of "if-then" statements is critical.

11. **The "if-then" statement is not to be identified with any particular type of relationship, causal, definitional, etc.** For instance, "If Jones is man, then he is a male" is true by **definition.** "If Jones' leg is broken, then he must have fallen" asserts a **causal** relationship between Jones' fall and his broken leg. "If the Pope is Catholic, then I am a Baptist" asserts a relationship between two

phenomena which is neither definitional nor causal (in any obvious sense). These are all examples, however, of an "if p then q" assertion. This is often referred to as stating a relationship of material implication between **p** and **q**, such that **p** implies **q**.

PROBLEM SET:

Let P be equated with "Jim's being president." Let E be equated with "Jim's being elected." Let H be equated with "Jim being happy."

Interpret the following "if-then" statements, putting them in proper form. Example: If Jim is elected, then he will be President. E → P

1. If Jim is President, then he is elected.
2. Jim's election is a sufficient condition for his being president.
3. Jim's election is a necessary condition for his being president.
4. Jim's being President is a sufficient condition for his being elected.
5. Jim's being president is a necessary condition for his being elected.
6. If Jim is elected, then he is president.
7. Jim's being elected implies that he is president.
8. Jim being elected entails that he is president.
9. Jim is president only if he is elected.
10. Jim is elected only if he is president.
11. Jim's being president entails that he is happy.
12. If Jim's being elected is a sufficient condition for his being happy, then Jim is president.
13. If Jim's being elected entails that he is president, then Jim is happy.
14. If Jim is happy only if he is elected, then Jim is president.
15. If Jim is president only if he is happy, then Jim is elected.

CHAPTER 5: TRUTH TABLES AND "IF-THEN" STATEMENTS

A. Truth Tables and the Meaning of P → Q.

Another way of defining the meaning of "If p then q" is through the use of truth tables. If we consider the statement variables p and q we can see that there are four and **only four** logical possibilities with respect to the truth and falsehood of these two variables taken together. (1) Both p and q may be true. (2) p may be true and q may be false. (3) p may be false and q may be true. (4) Both p and q may be false. Using the number 1 to equal true and 0 to equal false, these four logical possibilities are expressed below in figure 5.1

Figure 5.1

	p	q
1	1	1
2	1	0
3	0	1
4	0	0

The assertion that **p** entails **q** can best be understood as a denial of one of these four logical possibilities namely logical possibility number two. If p then q, does not assert that p is true. It does not assert that q is true. It merely asserts that on those occasions when p is true, that q has to be true also. In other words it can never be the case that p is true simultaneously with q being false. Or it is not the case that p is true and that q is false. We will illustrate this below with truth tables using what is known as a tilde (~) for a negation sign and the operand (**&**) for a conjunction sign. First, we need to understand the meaning of conjunction.

The meaning of **&** or conjunction is something that we all intuitively understand but since it is so critical to being able to grasp the essentials of a logical system it may be well to take time here to go over it.

Let us take the simple statement, "I have a cat and a dog." What conditions must exist for this **whole** statement to be true? Two conditions. "I have a cat" must be true. But "I have a dog" must also be true. If **either** statement is false, then the statement that "I have a cat and a dog" is also false. Using the capital letter C to equal "I have a cat" and the capital letter D to equal "I have a dog," we may illustrate the statement "I have a cat and a dog" as follows: C & D. With respect to the truth and falsehood of the two statements (1) "I have a cat" and (2) "I have a dog" there are four logical possibilities. These are illustrated by the truth table given below in figure 5.2.

Figure 5.2

	C	D
1	1	1
2	1	0
3	0	1
4	0	0

(1) They both may be true. (2) It is true that I have a cat but false that I have a dog. (3) It is false that I have a cat but true that I have a dog. (4) Both statements are false. Under what condition is the assertion that "I have a cat and a dog" true? Obviously only under the conditions of logical possibility number one. This is depicted below in figure 5.3

Figure 5.3

	C	D	(C & D)
1	1	1	1
2	1	0	0
3	0	1	0
4	0	0	0

"I have a cat and a dog" is true if **both** parts of
the conjunction are true. Otherwise it is false.

How might we construct a truth table for ~C and ~D? This is easy.
The negation of a true statement is always false and the negation of a
false statement is always true. Thus if it is true that I have a cat (C),
then the statement which denies this truth (~C) must be false. Thus, the
truth table for ~C will be just the opposite of the truth table for C and
the truth table for ~D will be just the opposite of the truth table for D.
The results of this are depicted in figure 5.4 below.

Figure 5.4

	C	D	~C	~D	(C & D)	(~C & ~D)
1	1	1	0	0	1	0
2	1	0	0	1	0	0
3	0	1	1	0	0	0
4	0	0	1	1	0	1

We have already constructed in figure 5.3 above a truth table for the
conjunction of (C & D). We need, for reasons that will become
obvious immediately below to construct a truth table for the
conjunction of (C & ~D). Let us take this one step at a time using the
order we have become accustomed to. The truth table for C and ~D is
given below in figure 5.5.

Figure 5.5

	C	~D
1	1	0
2	1	1
3	0	0
4	0	1

What shall we say about the truth table for the conjunction of (C & ~D)? Remember, a conjunction is true only where <u>both</u> parts of the conjunct are true. Therefore, only logical possibility number 2 in figure 5.5 above satisfies this condition. This is shown in figure 5.6 given below.

Figure 5.6

	C	~D	(C & ~D)
1	1	0	0
2	1	1	1
3	0	0	0
4	0	1	0

Lets see if we can translate the truth table for (C & ~D) into what will admittedly be somewhat convoluted English. Logical possibility number 1 in figure 5.6 indicates a state where it is true that I have a cat and false that I do not have a dog. Logical possibility number 2 indicates a state where it is true that I have a cat and true that I do not have a dog. Logical possibility number 3 indicates a state where it is false that I have a cat and false that I do not have a dog. Logical possibility number 4 indicates a state where it is false that I have a cat and true that I do not have a dog.

If you have not fully grasped everything up to this point, it may be useful to stop here and review the above material before proceeding.

Now what happens if we negate the whole truth table for (C & ~D)? Remember, the negation of a false statement is true and the negation of a true statement is false. Therefore, we may expect that the truth table for ~(C & ~D) will be exactly the opposite of the truth table for (C & ~D). This is depicted in 5.7 below.

Figure 5.7

	(C & ~D)	~(C & ~D)
1	0	1
2	1	0
3	0	1
4	0	1

Lets put together what we have developed so far. Figure 5.8 combines the elements of figures 5. 1-5.7 plus a truth table for p → q.

Figure 5.8

	C	D	~C	~D	(C & D)	(C & ~D)	~(C & ~D)	p → q
1.	1	1	0	0	1	0	1	1
2.	1	0	0	1	0	1	0	0
3.	0	1	1	0	0	0	1	1
4.	0	0	1	1	0	0	1	1

Notice that the truth tables for ~(C & ~D) and p entails q are identical.

If things are not fully clear up to this point you may want to review the development of our discussion step by step.

Otherwise, let us concentrate on logical possibility number 2 across the board. This is shown in Figure 5.9. This possibility depicts C as true and D as false. Therefore, (C & D) is false. But since ~D is true, (C & ~D) is true. But since (C & ~D) is true, ~(C & ~D) must be false.

Figure 5.9

Logical possibility number 2

C	D	~C	~D	(C & D)	(C & ~D)	~(C & ~D)	p → q
1	0	0	1	0	1	0	0

We are now in a position to understand the meaning of "if-then statements" in symbolic logic in terms of truth tables. Suppose I assert that "If I own a cat, then I own a dog." (C → D). What is being asserted here? I am not telling you that I own both a cat and a dog. I am not telling you that I own either a cat or a dog, or that I don't own either a cat or a dog. I am simply excluding one possibility, namely that situation in which I own a cat and do not own a dog. In other words, I am excluding what we have been designating as logical possibility number 2. In other words, I am saying that logical possibility number 2 (C & ~D) is false or ~(C & ~D) is true. As figures 5.8 and 5.9 above indicate p → q is logically equivalent to ~(p & ~q). ~(p & ~q) is simply the denial of logical possibility number 2.

Figure 5.10

C	D	(C →D)
1	1	1
1	0	0
0	1	1
0	0	1

Let us attempt to summarize the immediately proceeding discussion. The statement "if I have a cat, then I have a dog C → D" is really not an affirmation about my having either a cat or a dog. It is a **conditional** if /then statement. It is better understood as a denial of the condition where I have a cat simultaneously with my **not** having a dog. It denies (C & ~D. This is **all** that it does within the logical system with which we are working. Every other logical possibility is treated as true.

B. Some Odd Implications of the Meaning of p → q

We now need to consider some further examples of the meaning "if-then" statements.* These examples will, I hope, both clarify our meaning and illustrate some odd features of that meaning in the context of the logical system under consideration.

These features may be summarized as follows: (1) the meaning of p → q asserts a relationship between p & q which is "reasonable" only in **a very restricted sense**; (2) if the p statement is false, then it entails anything in the universe; (3) if the q statement is true, then it is entailed by anything in the universe.

Let us for the sake of the argument treat the following statements as true. "The author of this text is a college professor (P)." "The chemical properties of water are two parts hydrogen and one part oxygen (W)." Since, the two statements given above are true it is logically correct to assert that P → W or, "If the author of this text is a college professor then the chemical properties of water are two parts hydrogen and one part water." This is the case since it does not violate the condition where the antecedent (the p statement) is true and the consequent (the q statement) is false.

Let us also assume that the following statements are false. "The author of this text is an insurance agent (I)" and "2 + 2= 5 (F)." Since these two statements are false it is logically correct to assert that I → F or "The author of this text being an insurance agent, entails that 2 + 2 equals five." This is the case since like the example above it does not violate the condition where the antecedent is true and the consequent is false.

Let us assume the truth of the following: "Most art critics consider Michelangelo's Sistine Chapel to be a great work of art (A)." "There was a world wide depression which began in 1929 (D)." From these two statements we construct the following: "Most art critics consider Michelangelo's Sistine Chapel to be a great work of art (A) only if there was a world wide depression which began in 1929 (D), A→D." How can this statement be true? Once again, it is true because it fails

---

*As you recall from our discussion of ordinary English and "if-then" statements in chapter 4, there are a number of ways of stating "if p then q." You may want to review this section before proceeding.

to violate that condition where the antecedent is true and consequent is false.

Let us consider two more statements. "The Christian doctrine of the Trinity (T) being true is a sufficient condition for the truth of the Japanese attack on Pearl Harbor being on December 7, 1941, (J). T → J." "The Christian doctrine of the Trinity (T) being false is a sufficient condition for the Japanese attack on Pearl Harbor being on December 7, 1941 (J), T → J." Both these statements are true since they also fail to violate that one condition where the antecedent is true and the consequent is false.

Let us consider some implications of our logical system which is illustrated by the strange examples give above.

1. The Restricted Sense of "Rational" of the Logical System

Based on what we know about vocational occupations, the chemical properties of water, mathematics, art, military history and theology the relationships asserted above all appear to be irrational.

This is because we have a tendency to assume that there is some level of causal or other meaningful relationship being asserted than the restricted relationship designated by a specific logical system. There is of course no "causal" relationship between the author of this text being a college professor and the chemical properties of water. The truth or falsehood of the doctrine of the Trinity is not linked to the Japanese attack on Pearl Harbor. The falsehood of "2 + 2=5" has nothing to do with the falsehood of "The author of this text is an insurance agent."

If we interpret these statements as asserting some **broader** connection between these phenomenon, they will necessarily appear irrational. If we understand the restricted sense that is intended and only if we understand that restricted sense, will they be seen as rational.

2. A False "p" Statement Entails Anything in the Universe

The **only** way p → q can be false is if the p statement is true while the q statement is false. This can **never** happen if the p statement is false. Thus, if "The author of this text is an insurance agent (I)" is false, then it will entail any and everything. This is illustrated by the example given below.

"The author of this text is an insurance agent (I) entails that 4 + 4 equals eight (E), I → E." "The author of this text is an insurance agent (I) entails that 4 + 4 equals nine (N), I → N."

3. A True Statement is Entailed by Anything in the Universe

Consider the following examples:

"Most art critic's consider Michelangelo's Sistine Chapel to be a great work of art (A) only if there was a world wide depression which began in 1929 (D) or A → D." "If the author of this text is Abraham Lincoln (L), there was a world wide depression which began in 1929 (D) or L → D." "The author of this text being a hippopotamus (H) is a sufficient condition for there being a world wide depression which began in 1929 (D) or H → D." Thus, if the consequent is true, then it is entailed by true states of affairs, false states of affairs, descriptions of ridiculous states of affairs, etc. Why is this so? Because if the consequent is true, then we can never have a false consequent at the same time we have a true antecedent.

C. Truth Tables and Multiple Variables

Up to this point we have been developing truth tables which use two variables **p** and **q**, or **C** and **D**. With two variables there are only four logical possibilities. As we shall see the number of possibilities doubles every time we add a variable so that three variables gives us eight logical possibilities, four gives us sixteen, five thirty two, etc.

One standard way of setting up the truth tables (and the one we will use throughout) so as to insure that all logical possibilities will be represented is as follows: The first vertical column on the far right should start with true, false; true, false or 1, 0; 1, 0; until the column is filled. The second column from the right should double the number of true designations to appear first as in 1, 1; 0, 0; 1, 1; 0, 0; . The third column should double again the number of designations for true followed by double the number of designations for false that appeared in column two i.e. 1, 1, 1, 1; 0, 0, 0, 0 until all of the horizontal columns are filled. Thus, a truth table with four variables would have four horizontal columns and sixteen vertical rows. The fourth column

Logic of the Computing Sciences

on the far right would be filled in by alternating 1, 0 until all sixteen rows were filled. The 3rd column would be filled in by alternating 1, 1; 0, 0; The second by alternating 1, 1, 1, 1; 0, 0, 0, 0; the first by alternating 1, 1, 1, 1, 1, 1, 1, 1; 0, 0, 0, 0, 0, 0, 0, 0 until all sixteen rows are filled. This is shown below

Figure 5.11

	p	q	r	s
(1)	1	1	1	1
(2)	1	1	1	0
(3)	1	1	0	1
(4)	1	1	0	0
(5)	1	0	1	1
(6)	1	0	1	0
(7)	1	0	0	1
(8)	1	0	0	0
(9)	0	1	1	1
(10)	0	1	1	0
(11)	0	1	0	1
(12)	0	1	0	0
(13)	0	0	1	1
(14)	0	0	1	0
(15)	0	0	0	1
(16)	0	0	0	0

D. "If-Then" Statements and the Rules of Inference

We have referred above to the relationship between $p \rightarrow q$ as one of material implication. Understanding of material implication can be extremely useful in understanding the meaning of rules of inference. Let us take for instance Modus Ponens. Modus Ponens asserts that **if** the statement $p \rightarrow q$ is true **and** if the statement p is true, then q must also be true. This is depicted in Figure #12 below.

Figure 5.12

p	→	q
p		
∴		q

Another way of putting this is to assert that there is a relationship of material implication between the premises (p → q) and p and the conclusion q. We can show this in the form of truth tables. We will do this in a step by step process starting with figure 5.13 in order to enhance clarity.

Figure 5.13

	p	q
(1)	1	1
(2)	1	0
(3)	0	1
(4)	0	0

Figure 5.13 simply depicts the four logical possibilities regarding the truth and falsehood of p and q combined. Now let us put this together with (p → q) in figure 5. 14.

Figure 5.14

	p	q	(p → q)
(1)	1	1	1
(2)	1	0	0
(3)	0	1	1
(4)	0	0	1

We see that p → q is false <u>only</u> in logical possibility number 2. We observe that p is true for logical possibilities number 1 and 2 and false for 3 and 4. Let us now add a truth table for the conjunction of [(p → q) & p] in figure 5. 15.

Logic of the Computing Sciences

Figure 5.15

	p	q	(p → q)	(p → q) & p
(1)	1	1	1	1
(2)	1	0	0	0
(3)	0	1	1	0
(4)	0	0	1	0

We see from this truth table that there is only one case where both (p → q) and p are true. This is logical possibility number 1.

What Modus Ponens or M.P. asserts is that if (p → q) is true and if p is true, then q must be true. In other words the conjunction of [(p → q) & p] → q.

Let us add this to our truth table in figure 5. 16 to see if it is so.

Figure 5.16

	p	q	p →q	[(p	→	q)	&	p]	→	q
(1)	1	1	1	1			1	1	1	1
(2)	1	0	0	0			0	1	1	0
(3)	0	1	1	1			0	0	1	1
(4)	0	0	1	1			0	0	1	0

As can be shown from the above, the truth table for (p → q and p) entails q, is all 1's. This demonstrates that we can never have an instance in which [(p → q) & p] are true while q is false. If you take each of the rules of inference we will study and do a truth table on them you will get the same result. This is what makes them **valid** rules of inference. But it is also important to notice precisely what is being proven. Modus Ponens and the other rules of inference do not demonstrate a causal relationship between the variables dealt with. Modus Ponens simply **denies** the logical possibility in which [(p → q) & p] are true and q is false.

Let us put this in a slightly different way   As we check out logical possibilities 1, 2, 3 and 4 we see that there is no case where (p → q) and p are both true simultaneously with q being false.   In our illustration we are treating the combination of (p → q) & p as the entire p statement. We are treating q as the a q statement.   Since we never have a situation where [(p →q) & p] is true while q is false, Modus Ponens is shown to be a valid argument form.

If you understand this argument, you need no longer take the author's word for the fact that M.P. is a valid argument form. You can demonstrate it for yourself.

For many, the above explanation will have been unnecessarily repetitious. Others may need to go over this section again. If you are in this latter group then take the time to review the material. Understanding basic concepts in the beginning will make your work much easier later on.

Frequently, it is helpful to look at the same phenomena in different ways. In the next two chapters we have diagrammed the meaning of (p & q), (p → q) and Modus Ponens using flow charts and algorithms. The use of algorithms is especially common in the development of computer programs.

PROBLEM SET:

Let C equal "I have a cat." Let D equal "I have a dog." Let M equal "I have a mouse." Let H equal "I have a horse."

1. Construct a truth table depicting all the logical possibilities with respect to having or not having a cat and a dog.

2. Construct a truth table depicting all the logical possibilities with respect to having or not having a cat, a dog, a mouse and a horse.

3. Assume that "If I have a cat, then I have a dog" is a true statement. Assume also that "I have a cat is a true statement." Show by truth tables that these statements entail the statement that "I have a dog."

CHAPTER 6:   FLOW CHARTS

Flow Charts are ways of diagramming the logical structure of simple arguments and have been used among other purposes for diagramming computer programs.   They are not used extensively in computer programming now, largely because they require so much space and thus are not useful in dealing with long, complicated programs with many different steps.

We will use them in introducing new principles throughout this text for pedagogical purposes, since a visual representation in a diagram is for many people the best way to comprehend a new concept. Furthermore, introducing logical principles through flow charts serves to illustrate the relationship between logic and the computing sciences. There are several symbols which are commonly used with flow charts which we will introduce at the beginning of this chapter.

Figure 6.1

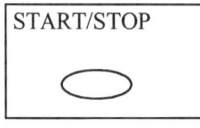

The oval is the symbol for terminal and/or interrupt (start, stop).

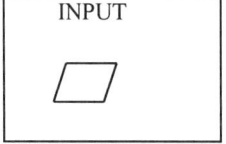

The parallelogram is the symbol for input and records the truth or falsehood of known variables provided by program.

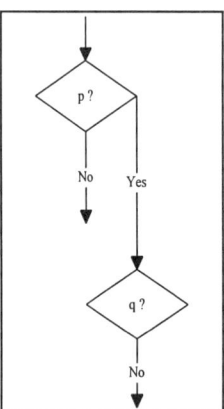

The diamond is the symbol for decision which directs the flow of traffic for the diagram.   It functions as the "then" statement.   For example, if the answer to the question "Is p true?" is "yes",   then   the   flow   chart   gives directions to either a further question or a particular output.   If the answer to the question "Is p true?" is no, then the chart gives directions to a different question or output.

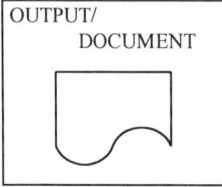

The scroll is the symbol for output (document) and contains the information which   would   be   printed   out   on   a computer program.

Figure 6.2

Flow Chart for p & q

Given that the value of p is known, and that the value of q is known; what can be known about the value of p & q?

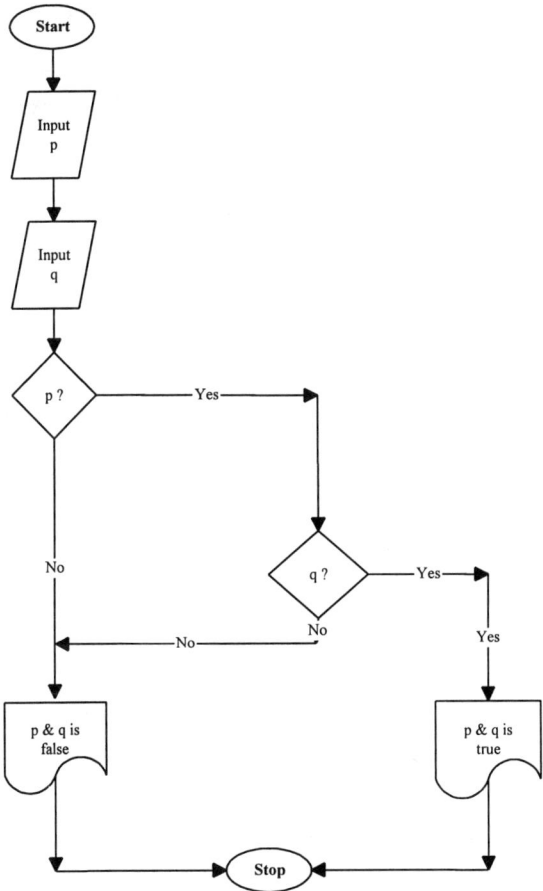

The explanation of the flow chart for (p & q) is simple but useful for our purposes. We know from simple analysis of ordinary English and our discussion of truth tables that (p & q) is true only if two conditions are met. p must be true. q must also be true. If either one of these conditions is not met, then (p & q) is false.

Thus, as our flow chart shows, if the answer to the question "Is p true?" is in the negative, then no further questions are necessary. It doesn't matter whether q is true or not, (p & q) will be false. If the answer to the question "Is p true?" is in the affirmative, then it becomes necessary to inquire about the truth value of q. If we have a negative here then we know that (p & q) is false. If the answer is affirmative then (p & q) is true.

Figure 6.3

Flow Chart for $p \to q$

Given that the value of p is known, and that the value of q is known; what can be known about the value of $p \to q$?

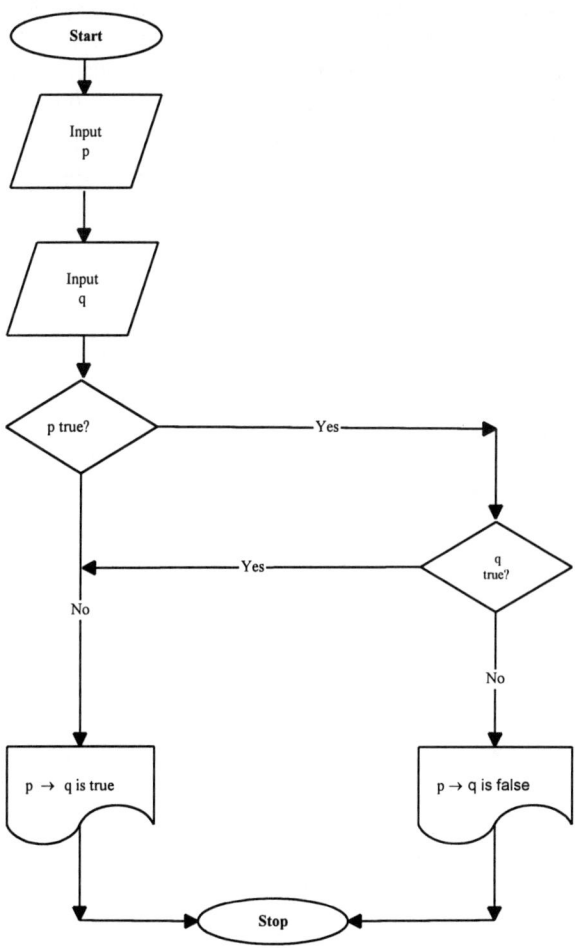

The explanation of the flow chart for p → q parallels that of our explanation in terms of truth tables. If the answer to question "is p true?" is no, then we proceed directly to the output statement which affirms that "p → q is true." This is the case because if p is false, then p → q will be true regardless of the truth value of q. Why is this so? Remember that p → q can be false only if p is true simultaneously with q being false. If p is false, then we can not have a true p and a false q. If p is true, then we must proceed to ask about the truth value of q in order to determine the truth of p → q. If q is true, then p → q is true and we proceed to the output statement which affirms this fact. If q is false, then we have a situation where p is true and q is false which makes p → q false. The output statement on the right indicates this fact.

Figure 6.4

Flow Chart for Modus Ponens
Given the values of p → q, and p what can be known about q?
[(p → q) & p] → q

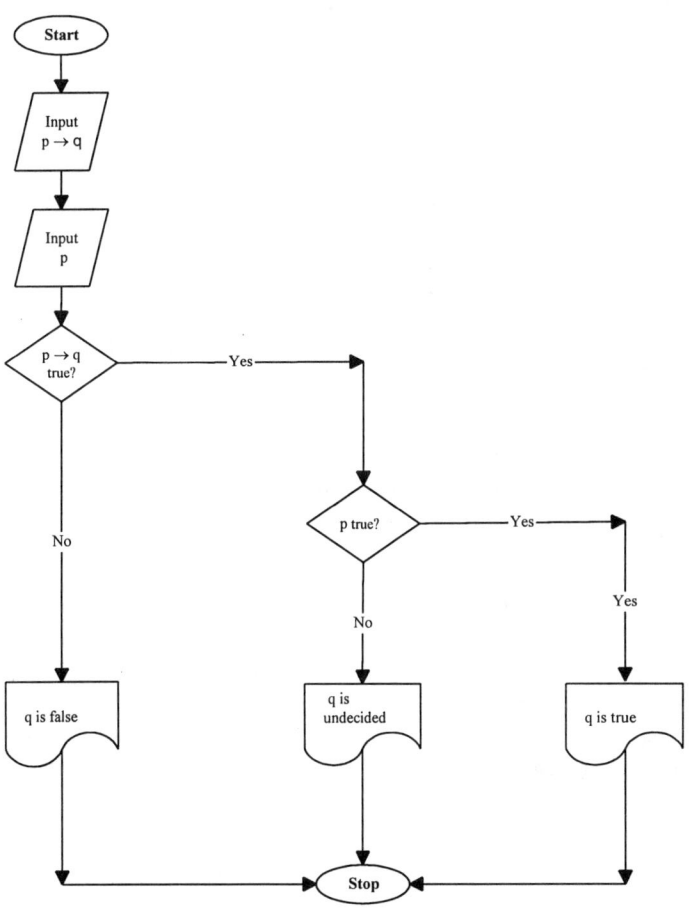

Our explanation for the flow chart for modus ponens is only slightly more complicated than the ones we have considered previously. The first question we ask concerns the truth value of
$p \rightarrow q$. If $p \rightarrow q$ is false we know that q is false, since $p \rightarrow q$ can be false only if p is true while q is false. If $p \rightarrow q$ is true, then we are left with only three possibilities since what we have called logical possibility number 2 has been excluded.

Figure 6.5

	p	$\rightarrow$	q
1.	1	1	1
2.	1	0	0
3.	0	1	1
4.	0	1	0

Next, the flow chart directs us to ask if p is true. If the answer is no, then logical possibility number 1 is also excluded.

Figure 6.6

	p	$\rightarrow$	q
1.	1	1	1
2.	1	0	0
3.	0	1	1
4.	0	1	0

But this leaves us with either logical possibilities number 3 and number 4. number 3 shows q as true, while number 4 shows q as false. Thus, since q can be either true or false, the flow chart goes to the output statement which says that "the value of q is undecided."

If the answer to the statement is p true is yes, then rather than excluding logical possibility number 1, we exclude possibilities number 3 and number 4 and are left only with number 1. Logical possibility number 1 has q as true. Therefore, our flow chart goes to the output statement "q is true."

Figure 6.7

	p	→	q
1.	1	1	1
2.	1	0	0
3.	0	1	1
4.	0	1	0

Below is an alternative flow chart for p → q which is also correct. Explain in your own words why this is the case.

Figure 6.8

Flow Chart for p → q

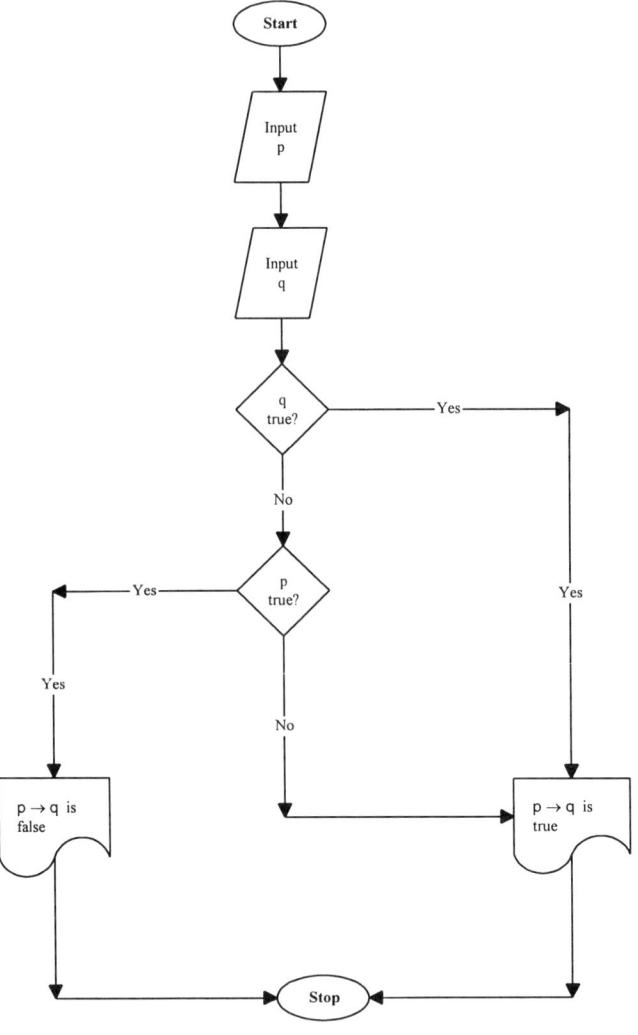

CHAPTER 7: ALGORITHMS

As we have indicated earlier, an algorithm is a set of procedures for performing certain operations which enable us to achieve a particular result.

The development of algorithms is one of the first steps commonly taken in the development of computer programs. The reader will readily observe that algorithm for (p & q), (p → q) and modus ponens [(p → q) & p] → q provide alternatives to ordinary English explanations, truth tables and flow charts. Below are given algorithms for all three of these relationships.

Figure 7.1

Algorithm For p & q

Given the value of p and also given the value of q, what can be known about the value of (p & q)?

1. Input value of p.
2. Input value of q .
3. If p is false,
     (a) (p & q) is false. **
     (b) Stop
4. If q is false,
     (a) (p & q) is false.
     (b) Stop.
5. (p & q) is true.
6. Stop.

If you have understood the chapter on flow charts, then this explanation will be repetitive. Since repetition is sometimes necessary or useful if we are to understand basic concepts, we offer a detailed explanation here.

Our first algorithm is for (p & q). We know that (p & q) is true, if and only if, p is true and q is also true. A false p makes this impossible. Therefore, in step three we are told that if p is false, to output "(p & q) is false." The algorithm then is instructed to stop, since consideration of the value of q is unnecessary to determine the outcome.

If p is not false, (then it must be true) and the algorithm **automatically** proceeds to number four. If q is false, then we know that (p & q) is false and our algorithm gives this output and proceeds no further.

If q is not false (then it must be true) and the algorithm proceeds to step number five and gives the output "(p & q) is true." Having completed its task we proceed to step six and stop.

** Another way of symbolizing "(p & q) is false," is "~(p & q)".

Figure 7.2

Algorithm For p → q

Given the value of p and also given the value of q, what can be known about the value of p → q?

1. Input value of p.
2. Input value of q.
3. If value of p is false,
    (a) (p → q) is true.
    (b) Stop.
4. If value of q is true
    (a) (p → q) is true.
    (b) Stop.
5. (p → q) is false.
6. Stop.

As our illustration indicates, p → q stands for "if p then q". This is identical with ~(p & ~q) which we have previously shown by truth tables. As step three shows, if the value of p is false, then the output "p → q is true" is appropriate. This is the case because as we have observed previously, if the antecedent (the "p" statement) is false then p → q is true regardless of the truth or falsehood of the q statement. Of course, if p is not false, then it must be true and we move to step four. If at this point we discover the q statement to be true, we have p → q and the output "p → q is true" is still appropriate. Only if q is false do we move to step five and output "p → q is false" since we have in this instance simultaneously a true p and a false q.. Having finished our task of analysis we move to step six and stop.

Figure 7.3

Algorithm For Modus Ponens
$[(p \rightarrow q) \, \& \, p] \rightarrow q$

Given the value of $p \rightarrow q$ and also given the value of p, what can be known about the value of q?

1. Input value of $p \rightarrow q$.
2. Input value of p.
3. If $p \rightarrow q$ is false,
    (a) q is false.
    (b) Stop.
4. If p is false,
    (a) q is undecided.
    (b) Stop.
5. q is true.
6. Stop.

The algorithm for modus ponens or $[(p \rightarrow q) \, \& \, p] \rightarrow q$ is a little more complicated since we have three possibilities "q is true," "q is false," and "q is undecided."

In step three, if $p \rightarrow q$ is false, then we know that q is false and our algorithm stops. This is true, of course, because $p \rightarrow q$ can be false only if the antecedent (the p statement) is true while the consequent (the q statement) is false. Compare logical possibility number 2 in Figure 7.4 below.

If $p \rightarrow q$ is not false, then it must be true and we move on to step number four. Four asks, "is p false?" If p is false then, as we indicated in our discussion of flow charts, we exclude logical possibility number 1 and are left with logical possibility number 3 (p false and q true) which makes q true or logical possibility number 4 (p false and q false) which makes q false. Since we do not have information to distinguish between logical possibilities number 3 and number 4, we output "the value of q is undecided."

If p is not false, then it must be true and we move on to step number five. Since $p \rightarrow q$ is true, we have excluded logical possibility number 2. Since p is true, we have excluded number 3 and number 4. This leaves us with number 1 in which q is true. Thus, we have the output denoted in step number five in figure 7.3.

Figure 7.4

	p	q	p → q
1	1	1	1
2	1	0	0
3	0	1	1
4	0	0	1

You may want to review the preceding, comparing it with figure 7.4.

The reader has now been introduced to the concepts of truth tables, flow charts and algorithms. We will develop these concepts further as we proceed. What should be clear is that what we are dealing with are different ways of expressing the same basic concept. Thus, if the reader has fully grasped the meaning of one mode of presentation, then he or she will have grasped in principle, the meaning of all the others.

This may be perceived as simple repetition. There are, however, several advantages to this approach. As we have observed previously, looking at the concepts from different points of view enhances understanding. Furthermore, what we are trying to do in this text is call attention to and demonstrate the relationship between logical systems, algorithmic thinking and computer programs. This approach will be used throughout and is designed to be of use not only for those who intend to do work in computing sciences, but also for those who seek an understanding of the logic of program systems at the most basic level.

Thus, if you haven't fully understood the previous chapters you may want to review them now. For some it may be helpful to move on to the next chapter and return to these earlier chapters later.

CHAPTER 8: MODUS PONENS

We have already introduced the conceptual basis of the logical principle known as Modus Ponens. Below you will find two sets of problems. The first set is already cast in symbolic form. The second set contains word problems which must be translated into symbolic form before they can be solved. Generally speaking, they are arranged in ascending order of difficulty so that number 1 is relatively easy, while numbers 8, 9, and 10 are much more difficult. Before you attempt the word problems, you may especially want to review chapter IV which specifically treats the varied ways in which "if-then" statements may appear in ordinary English.

It will be helpful to consider some basic rules of notation and principles of problem solving before attempting to solve the problems in this set. We have alluded to these rules and principles earlier (see chapter IV). Since they are applicable throughout our study, however, we will review them here.

Numbering of Steps

Modus Ponens takes the following form as in figure 8.1.

Figure 8.1

P	→	Q
P		
∴ Q		

It does not take the form shown in figure 8.2.

Figure 8.2

P		
P	→	Q
∴ Q		

Of course, the second form can be shown by truth tables to be a valid argument form. It is, however, different from the first one. We will be introducing a system which has twenty-three basic rules. It would be possible to develop a system with thirty, a hundred, or more rules. Most of them would be only slight variations of the original twenty three. This would, obviously, complicate our task unnecessarily.

It is important that our notation indicate clearly which premises and which rules are being used to achieve a particular result. This is a requirement for clear thinking. Those who have developed computer programs know that this is essential. Computers do not tolerate even apparently trivial errors very well.

Let us symbolize: "I have a cat. If I have a cat, then I have a dog. Therefore, I have a dog."

Figure 8.3

1.	C			∴ D
2.	C	→	D	

The correct procedure for solving this problem is depicted below in figure 8.4.

Figure 8.4

Correct				Incorrect		
#1				#2		

1.	C	∴ D		1.	C	∴ D
2.	C → D			2.	C → D	
3.	D	2,1 M.P.		3.	D	1,2 M.P.

Why is example #1 in figure 8.4 correct and example #2 wrong? Because example # 1 indicates by giving 2,1 M.P. in step three that the form Modus Ponens is correctly used by putting the entailment proposition first. This is indicated in figure 8.8

Figure 8.5

p	→	q		2	C	→	D
p				1	C		
∴	q			∴	D		

Example #2 in figure 8.4, however, by giving 1,2 M.P. in step three indicates the form shown in figure 8.6.

Figure 8.6

p				1	C		
p	→	q		2	C	→	D
∴	q			∴	D		

This, as we have indicated earlier, is a variation on Modus Ponens but is not Modus Ponens itself. Therefore, although it is not logically invalid, it is incorrect according to the standards of notation we have adopted.

Starting with the Conclusion

The importance of the use of proper notation and starting with the conclusion will become more obvious as we introduce more principles and problems with numerous premises. The example of the problem given below will perhaps be sufficient for the present.

Figure 8.7

1.	C			∴ G
2.	C	→	D	
3.	E	→	F	
4.	F	→	G	
5.	D	→	E	

Our task is to take the premises given in 1-5 and prove G. How may we best proceed? First, we observe that our conclusion calls for us to prove G. Secondly, we scan our premises and look for the presence of G. We discover it in premise #4 which affirms F → G. We conclude that if we had an F, we could combine it with premise #4 and Modus Ponens to prove G.

Since a scan of our premises indicates that we don't have an F our new problem becomes discovering a proof for F. Once again we scan our premises. This time we discover in premise #3 that we have the affirmation of E → F. We conclude that if we had an E, we could combine it with premise #3 and Modus Ponens to prove F.

Since we don't have an E, our new problem is to develop a proof for E. We scan our premises again and discover that in premise #5 we have the affirmation of D → E. We conclude that if we had a D, we could combine it with premise #5 and Modus Ponens to prove E.

Since we don't have a D, the proof of D now becomes our new problem. We scan our premises once more and discover the affirmation of C → D in premise #2. We conclude that if we had a C then we could combine it with premise #2 and Modus Ponens to prove D. Previous scans have doubtless already turned up the fact that we do have the existence of C affirmed in premise #1.

Now we know how to prove G which is our original goal. We will prove D, E, F and G in that order. This is shown below in figure 8.8.

Figure 8.8

1.	Ⓒ			∴ G
2.	C	→	D 2	
3.	E	→	F	
4.	F	→	G	
5.	D	→	E	
6.	D			2, 1 M.P.
7.	E			5, 6 M.P.
8.	F			3, 7 M.P.
9.	G			4, 8 M.P.

Punctuation in Symbolic Logic: Parenthesis and Brackets

In this text we will use the following conventions to indicate ever more inclusive breaks:

();   [   ];   {   };   〈   〉.

Proper punctuation is critical in any communication in which precision is demanded. When you are translating word problems into symbolic form it will be extremely important to observe the English punctuation since commas, semi-colons, and colons will be used to indicate logical breaks.

Perhaps the best way to indicate the way in which these punctuation marks may function is by the use of a series of arguments all of which are proper uses of Modus Ponens.

Figure 8.9

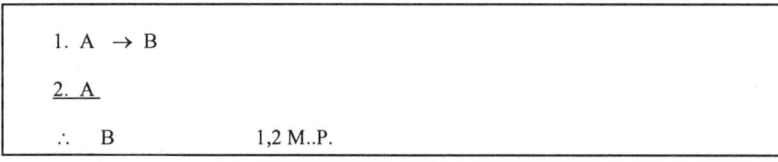

1. A → B

2. A

∴   B   1,2 M..P.

Figure 8.10

1. $(A \rightarrow B) \rightarrow (C \rightarrow D)$

2. $(A \rightarrow B)$

∴ $C \rightarrow D$ 1, 2 M..P.

Figure 8.11

1. $[(A \rightarrow B) \rightarrow (C \rightarrow D)] \rightarrow [(E \rightarrow F) \rightarrow (G \rightarrow H)]$

2. $[(A \rightarrow B) \rightarrow (C \rightarrow D)]$

∴ $(E \rightarrow F) \rightarrow (G \rightarrow H)$ 1,2   M..P.

Figure 8.12

1. $\{[(A \rightarrow B) \rightarrow (C \rightarrow D)] \rightarrow [(E \rightarrow F) \rightarrow (G \rightarrow H)]\} \rightarrow (I \rightarrow J)$

2. $\{[(A \rightarrow B) \rightarrow (C \rightarrow D)] \rightarrow [(E \rightarrow F) \rightarrow (G \rightarrow H)]\}$

∴ $(I \rightarrow J)$ 1,2 M..P.

Figure 8.13

1. $\{[(A \rightarrow B) \rightarrow (C \rightarrow D)] \rightarrow [(E \rightarrow F) \rightarrow (G \rightarrow H)]\} \rightarrow (I \rightarrow J) \rightarrow K$

2. $\{[(A \rightarrow B) \rightarrow (C \rightarrow D)] \rightarrow [(E \rightarrow F) \rightarrow (G \rightarrow H)]\} \rightarrow (I \rightarrow J)$

∴ K   1,2 M . P.

There are two important things to observe from these examples. The first thing to observe is that although the premises in examples Figures 8.10-8.13 are progressively more complicated than figure 8.9, that the logic is equally simple in all of the arguments. It is only Modus Ponens. In figure 8.13 we used K as a simple q statement. We might have used a q statement which was a mile long. The argument proving this mile long statement would still be only Modus Ponens.

Being able to "see" the <u>simple</u> <u>argument</u> <u>form</u> available for use in a long and complicated set of premises is the key to solving all problems in symbolic logic. It is also the key to solving a lot of problems (although not all) in real life.

Being able to translate complicated arguments from ordinary English into symbolic form is the first step in simplifying what at first glance may appear to be an impossibly complex argument.

This leads us to the second important thing to be observed from the examples given above. It is essential if we are to detect the simple argument form available to us in the premises, that we interpret correctly the punctuation marks which indicate which complex of symbols is the p statement and which is the q statement.

In our consideration of rules of notation and principles of problem solving we have stressed three things: (1) **proper numbering of steps; (2) solving problems by starting with the conclusion and working backward; and (3) careful consideration of punctuation in order to discern the simple argument form in the midst of a seemingly complex set of premises.** Let us put these lessons together in the solution of the following problem.

Figure 8.14

1.	$[(A \rightarrow B) \rightarrow (C \rightarrow D)] \rightarrow [(E \rightarrow F) \rightarrow (F \rightarrow G)]$	$\therefore$ G
2.	E	
3.	$E \rightarrow F$	
4.	A	
5.	$A \rightarrow [(A \rightarrow B) \rightarrow (C \rightarrow D)]$	

Let us start once again with our conclusion which is G. We observe that G appears in the first premise but we can't get at it immediately. We do see that <u>if</u> we could isolate $F \rightarrow G$ and <u>if</u> we had an F, then we could solve for G. Deriving F from 3, 2 M.P. is easy, but how can we get $F \rightarrow G$ out of our first premise? If we could isolate $(E \rightarrow F) \rightarrow (F \rightarrow G)$ we could use premise #3 and M.P. to derive $F \rightarrow G$. Now our problem is to derive $(E \rightarrow F) \rightarrow (F \rightarrow G)$. We see that if we had $(A \rightarrow B) \rightarrow (C \rightarrow D)$ we could use it along with M.P. to derive $(E \rightarrow F) \rightarrow (F \rightarrow G)$. We can derive $(A \rightarrow B) \rightarrow (C \rightarrow D)$ from 5, 4 M.P. Thus, we will start with this step and proceed to prove for G.

Figure 8.15

1.	$[(A \to B) \to (C \to D)] \to [(E \to F) \to (F \to G)]$		$\therefore$ G
2.	E		
3.	E $\to$ F		
4.	A		
5.	A $\to$ [(A $\to$ B) $\to$ (C $\to$ D)]		
6.	(A $\to$ B) $\to$ (C $\to$ D)	5, 4	M.P.
7.	(E $\to$ F) $\to$ (F $\to$ G)	1, 6	M.P.
8.	F $\to$ G	7, 3	M.P.
9.	F	3, 2	M.P.
10.	G	8, 9	M.P.

If you have understood our discussion up to this point you should be fully ready to tackle the problems included in this chapter.

PROBLEM SET: MODUS PONENS
Prove the validity of the following arguments.

1.

    1.   D $\therefore$ F
    2.   D $\to$ F

2.

    1.   D $\to$ F $\therefore$ X
    2.   F $\to$ X
    3.   D

3.

    1.   F $\to$ X $\therefore$ X
    2.   C $\to$ Y
    3.   Y $\to$ R
    4.   C
    5.   R $\to$ Z
    6.   Z $\to$ F

4.

1. $F \rightarrow [(C \rightarrow Y) \rightarrow (Y \rightarrow R)]$ $\quad\quad\quad\quad \therefore R$
2. C
3. F
4. $C \rightarrow Y$

5.

1. $[(D \rightarrow F) \rightarrow G]$ $\quad\quad\quad\quad \therefore H \rightarrow (I \rightarrow J)$
2. $[(D \rightarrow F) \rightarrow G] \rightarrow [H \rightarrow (I \rightarrow J)]$

6.

1. $A \rightarrow [(D \rightarrow F) \rightarrow G]$ $\quad\quad\quad \therefore J$
2. $\{A \rightarrow [(D \rightarrow F) \rightarrow G]\} \rightarrow \{B \rightarrow [H \rightarrow (I \rightarrow J)]\}$
3. B
4. H
5. I

7.

1. $[(A \rightarrow B) \rightarrow (C \rightarrow D)] \rightarrow \{(D \rightarrow E) \rightarrow [(F \rightarrow G) \rightarrow (H \rightarrow I]\}$
2. H $\quad\quad\quad\quad\quad\quad\quad\quad \therefore I$
3. $F \rightarrow G$
4. $D \rightarrow E$
5. J
6. $J \rightarrow [(A \rightarrow B) \rightarrow (C \rightarrow D)]$

8.

1. $(D \rightarrow E) \rightarrow [(F \rightarrow G) \rightarrow (H \rightarrow I)]$ $\quad\quad \therefore I$
2. $\{(D \rightarrow E) \rightarrow [(F \rightarrow G) \rightarrow (H \rightarrow I)]\} \rightarrow J$
3. $J \rightarrow (D \rightarrow E)$
4. $(D \rightarrow E) \rightarrow (F \rightarrow G)$
5. H

9.

1. $\{(D \rightarrow E) \rightarrow [(F \rightarrow G) \rightarrow (H \rightarrow I)]\} \rightarrow$ $\therefore$ N
     $\{J \rightarrow [(K \rightarrow L) \rightarrow (M \rightarrow N)]\}$
2. $(O \rightarrow P) \rightarrow \{(D \rightarrow E) \rightarrow [(F \rightarrow G) \rightarrow$
     $(H \rightarrow I)]\}$
3. $R \rightarrow [(S \rightarrow T) \rightarrow (O \rightarrow P)]$
4. $K \rightarrow L$
5. $\{(U \rightarrow V) \rightarrow [(X \rightarrow Y) \rightarrow (Y \rightarrow Z)]\}$
6.   M
7. $\{(U \rightarrow V) \rightarrow [(X \rightarrow Y) \rightarrow (Y \rightarrow Z)]\} \rightarrow J$
8.   R
9. $S \rightarrow T$

10.

1. $D \rightarrow \{(E \rightarrow F) \rightarrow [(G \rightarrow H) \rightarrow (I \rightarrow J)]\}$
2. $D \rightarrow \{(E \rightarrow F) \rightarrow [(G \rightarrow H) \rightarrow (I \rightarrow J)]\} \rightarrow (K \rightarrow L)$
3.   K
4. $L \rightarrow D$ $\therefore$ J
5.   I
6. $G \rightarrow H$
7. $P \rightarrow (E \rightarrow F)$
8.   P

1.   Jones is liable for damages only if he is negligent. Jones is liable for damages. Therefore, Jones is negligent. L, N

2.   Jones being negligent is a sufficient condition for him to be liable for damages. Jones was negligent. Therefore, Jones is liable for damages.
     N, L

3.   Jones being negligent is a necessary condition for his being liable for damages. Jones is liable for damages. Therefore, Jones is negligent. N, L

4.   Jones isn't liable for damages unless he was negligent. Jones is liable for damages. Therefore, Jones was negligent. L, N

5. If Jones being negligent is a sufficient condition for his being guilty, then if Jones is convicted he will have to pay a fine. Jones is negligent only if he is guilty. Jones will be convicted. Therefore, Jones will have to pay a fine. N, G, C, P.

6. If Jones is guilty only if he is negligent is a sufficient condition for his conviction entailing that he must pay a fine, then Jones is in trouble only if he was sleeping on the job. Jones not being guilty unless he was negligent is a sufficient condition for his conviction entailing that he must pay a fine. Jones is in trouble. Therefore, Jones was sleeping on the job. G, N, C, P, T, S.

7. Jones is guilty only if Jones' sleeping on the job is a necessary condition for his being negligent. Jones is negligent. Jones is guilty. Therefore, Jones was sleeping on the job. G, S, N.

8. Jones' being guilty is a sufficient condition for his paying a fine, only if Jones' sleeping on the job is a necessary condition for his being negligent. Jones is guilty only if he pays a fine. Jones is negligent. Therefore, Jones was sleeping on the job. G, P, S, N.

9. Jones is guilty if being liable for damages necessarily follows from sleeping on the job. Sleeping on the job is a sufficient condition for being liable for damages. Therefore, Jones is guilty. G, L, S.

10. Jones is guilty if it is the case that sleeping on the job is a sufficient condition for negligence. Negligence being a necessary condition for sleeping on the job is a necessary condition for Jones' paying a fine. Jones must pay a fine. Therefore, Jones is guilty. G, S, N, P.

CHAPTER 9: MODUS TOLLENS

Another valid argument form is known as Modus Tollens or M.T. It requires the use of the symbol for negation known as the "tilde" which was introduced earlier.   Thus, if I equals God is incomprehensible, ~I equals it is not the case that God is incomprehensible.

Modus Tollens or M.T. is symbolized as an argument form as shown in figure 9.1.

Figure 9.1

p → q
~ q

When we substitute capital letters for particular phrases we get an arguments as shown in figure 9.2-9.6.

Figure 9.2

1. If Jones is negligent then Jones is liable for damages.
2. It is not the case that Jones is liable for damages.
    Therefore, Jones is not negligent.

N   →   L
~ L
∴   ~N

Figure 9.3

1. If God is non-temporal, then his being is incomprehensible. (G → I)
2. God's being is not incomprehensible (~ I). Therefore, it is not the case that God is non-temporal (~ G).

G   →   I
~ I
∴   ~ G

Let me emphasize once again that a valid argument is one in which **if** the premises are true then the conclusion is **necessarily** true. This means, among other things, that a valid argument with false premises may have a false or even ridiculous conclusion. It also means that if the conclusion of a valid argument is shown empirically to be false that one or more of its premises must be false. For example:

Figure 9.4

1. If Lincoln was a man then Lincoln was a mammal (L → M).
2. Lincoln was not a mammal (~ M). Therefore, Lincoln was not a man ~ L.

The statement "Lincoln was not a man." is obviously false even through it is the conclusion of valid argument. This is because, the second premise "Lincoln was not a mammal." is false.

1.   L → M
2.   ~ M
∴ ~L

The above is a valid argument. (M.T.) The second premise, however, is false and the conclusion is also false.

It is useful at this point to remember the point made under #7 in chapter 4 regarding which portion of the statement is functioning as the p or q statement. In example figure 9.5 below we have an example of Modus Tollens.

Figure 9.5

1.	M → R
2.	~ R
	∴ ~M

M functions as the p statement. **R** functions as the q statement. The **negation** sign negates R or the q statement in the second premise and M or the p statement in the conclusion. But consider the argument given immediately below in figure 9.6.

Figure 9.6

~M → ~R
~M
∴ ~R

Despite the presence of negation signs, we have here an example of Modus Ponens not Modus Tollens. The negation sign is part of the p statement in the case of (~M) and is part of the q statement in the case of (~R).

Figure 9.7

Truth Tables For Modus Tollens

p	q	~p	~q	(p → q)	{[(p → q) & ~q]	→	~p
1	1	0	0	1	0	1	0
1	0	0	1	0	0	1	0
0	1	1	0	1	0	1	1
0	0	1	1	1	1	1	1

Figure 9.8

Flow Chart for Modus Tollens
Given the value of p → q, and q what can be known about p?

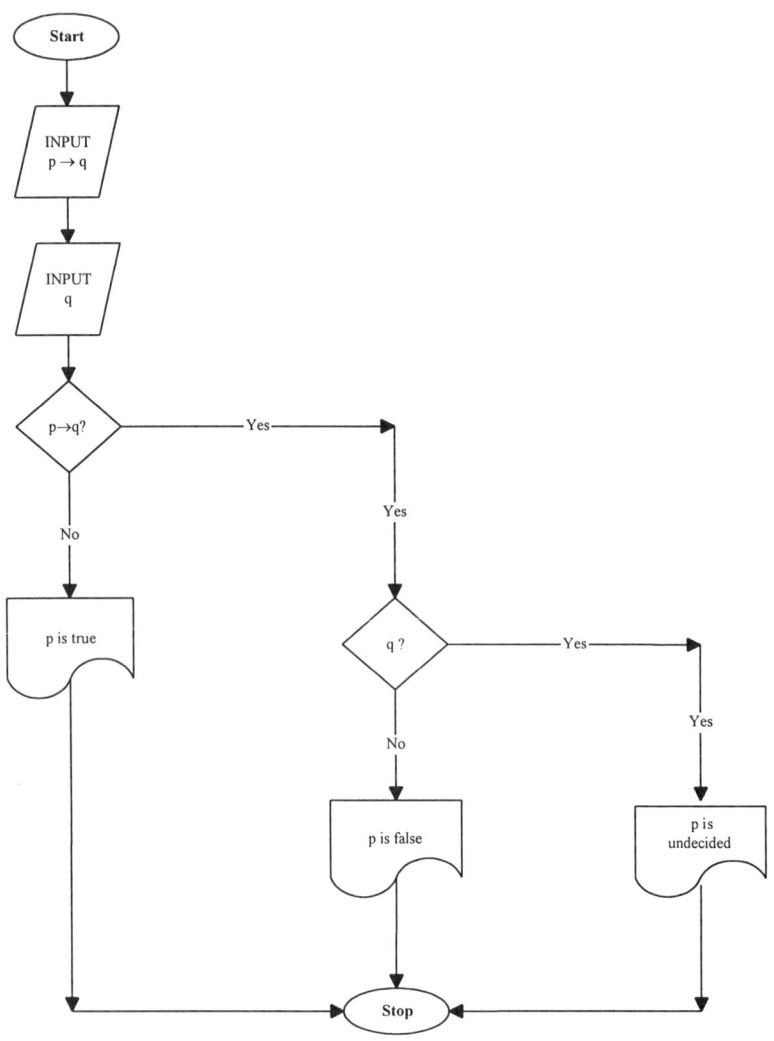

The explanation for the Modus Tollens flow chart parallels that for Modus Ponens. If p → q is false, then we know that p is true, since the only way p → q can be false is for p to be true while q is false. Our flow chart thus proceeds immediately to the output "p is true." If, however, p → q is true, then we must ask about the value of q. If p → q is true and q is also true, then the value of p is undecided since both possibility number 1 (which has p as true) and number 3 (which has p as false) are left open.(Compare truth table below.)

If, however, q is false, we exclude possibilities #1 and #3 in figure 9.9 below. The fact that p → q is true has already excluded #2. Thus, we are left with #4 which shows p as false.

Figure 9.9

	p	q	p	→	q
1.	1	1		1	
2.	1	0		0	
3.	0	1		1	
4.	0	0		1	

Figure 9.10

Algorithm For Modus Tollens
Given the value of p → q, and q what can be know about p?

1. Input value of p → q.
2. Input value of q.
3. If p → q is false,
       (a) p is true.
       (b) Stop.
4. If q is true,
       (a) p is undecided.
       (b) Stop.
5. p is false.
6. Stop.

The explanation for the Modus Tollens algorithm is identical with the explanation for the flow chart. This is not surprising since these are two different ways of presenting the same logical concept.

PROBLEM SET: MODUS TOLLENS
Prove the validity of the following arguments.

1.
> 1. ~B   ∴ ~A
> 2. A → B

2.
> 1. (A → B) → C   ∴ ~(A → B)
> 2. ~C

3.
> 1.   [(A → B) → (D → C)] → D   ∴ ~[(A → B) → (D → C)]
> 2.   ~D

4.
> 1. (A → B) → ~C   ∴ ~(A → B)
> 2. ~~C

5.
> 1. ~C   ∴ H
> 2. (A → B) → C
> 3. ~(A → B) → [(D → F) → (G → H)]
> 4. G
> 5. D → F

6.
> 1. ~C   ∴ ~(I → J)
> 2. (A → B) → C
> 3. ~(A → B) → ~[(D → F) → (G → ~H)]
> 4.   (I → J) → [(D → F) → (G → ~H)]

7.
> 1. (A → B) → [(C → D)→ (E → F)]   ∴ F
> 2. ~(G → F) → (A → B)
> 3. (G → F) → I
> 4. ~I
> 5. ~I → (C → D)

6. E

8.

1. $(A \to B) \to [(C \to D) \to (E \to F)]$ $\therefore \sim E$
2. $\sim(G \to F) \to (A \to B)$
3. $(G \to F) \to I$
4. $\sim I$
5. $\sim I \to (C \to D)$
6. $\sim F$

9.

1. $[(A \to B) \to (C \to D)] \to [\sim(E \to F) \to (\sim G \to H)]$ $\therefore$ H
2. $\sim I \to K$
3. $\sim L$
4. $I \to L$
5. $K \to [(A \to B) \to (C \to D)]$
6. $(E \to F) \to L$
7. $\sim J$
8. $G \to J$

10.

1. $A \to \{(B \to C) \to [(D \to F) \to (G \to H)]\}$ $\therefore$ G $\to$ H
2. $\sim I \to A$
3. $I \to J$
4. $\sim J$
5. $\sim K \to (B \to C)$
6. $K \to L$
7. $\sim L$
8. $\sim M \to (D \to F)$
9. $M \to N$
10. $\sim N$

1. Jones is liable for damages only if he is negligent. If Jones is a careful man then, Jones is not negligent. Jones is a careful man. Therefore, Jones is not liable for damages. L, N, C.

2. If Smith worked more than 40 hours, then he receives overtime pay. If Smith's pay check was less than 300 dollars, then Smith did not receive overtime pay. Smith's pay check was less than 300 dollars. Therefore, Smith did not work more than 40 hours. H, O, L.

3. If slum lords control the political structure of a community, then the legal services program will not have political support. Slum lords do have control of the political structure of the community. A legal services program can be

successful only if it has political support. Therefore, the legal services program will not be successful. C, P, S.

4. God is perfect only if he is omnipotent. The existence of evil is a sufficient condition for God not being omnipotent. Evil exists. Therefore, God is not perfect. P, O, E.

5. Man is truly human only if he is free. If man is free then he will do evil. God loves man only if man is truly human. God is perfect only if he loves man. God is perfect. Therefore, man will do evil. H, F, E, L, P.

6. If God is perfect then he is comprehensible. God is comprehensible only if logic can be used to understand his being. Logic being a useful discipline is a necessary condition for the possibility of logic being used to understand the being of God. Logic is not a useful discipline. Therefore, God is not perfect. P, C, L, U.

7. If God is perfect then he is not comprehensible. If God is not comprehensible then logic cannot be used to understand his being. Logic cannot be used to understand the being of God only if logic is not useful. It is not the case that logic is not useful. Therefore, God is not perfect. P, C, L, U.

8. If God is comprehensible, then he is not perfect. If God is not comprehensible, then reason is useful but limited. If reason is useful but limited, then it is not the case that logic's being used to understand the being of God is a necessary condition for logic's being useful. It is not the case that God is not perfect. Therefore, it is not the case that logic's being useful entails that logic can be used to understand the being of God.
    Let C = God is comprehensible.
    Let P = God is perfect.
    Let R = reason is useful but limited
    Let L = logic is used to understand the being of God
    Let U = logic is useful.

9. Mary is pregnant only if she is HCG positive (Human Coronic Gonadotropin). Mary is not HCG positive. Therefore, Mary is not pregnant. P, H (assume Mary last had intercourse a month ago and has no diseases)

10. If James Phipps (patient of William Jernner, 1749-1823) had not had cowpox, then he could have gotten smallpox. If Phipps had the antibodies for cowpox, then he could not get smallpox. Phipps had the antibodies for cowpox. Therefore, it is not the case that Phipps had not had cowpox. C, S, A (Let C = "had not had cowpox.")

CHAPTER 10: THE PRINCIPLES OF CONJUNCTION AND SIMPLIFICATION

Two of the most obvious logical principles we encounter in our study are those of conjunction (Conj.) and simplification (Simp.) Logically the two separate assertions (1) "Jones has a cat." (C), (2) "Jones has a dog." (D) are equivalent to the assertion (3) "Jones has a cat and a dog." If we use the symbol of the operand (**&**) to symbolize the combining or conjunction of these two assertions, they may be symbolized as follows: C & D. Conversely if statement (3) "Jones has a cat and a dog" is true then if follows logically that statement #1 is correct. "Jones has a cat." These principles are symbolized as follows:

Figure 10.1

Conjunction

p
q
∴   p & q

Simplification

p & q
∴   p

The proper use of these principles can be illustrated by the following problem.

Figure 10.2

Jones has a cat and a dog. Therefore, Jones
has a cat.

1.	C & D	∴ C	
2.	C		1, Simp.

Figure 10.3

If Jones has a cat and a dog, then Jones has a
hamster and a gerbil. Jones has a dog. Jones
has a cat. Therefore, Jones has a hamster.

1.	(C & D) & (H & G)	∴ H	
2.	D		
3.	C		
4.	C & D	3,2	Conj.
5.	H & G	1,4	M.P.
6.	H	5,	Simp.

A truth table, algorithm and flow chart for the principle of
simplification are given below.

Figure 10.4

Truth Table For Simplification

p	q	(p & q)	→	p
1	1	1	1	1
1	0	0	1	1
0	1	0	1	0
0	0	0	1	0

Figure 10.5

Flow Chart for Simplification

Given that the value of p & q is known, what can be known about the value of p?

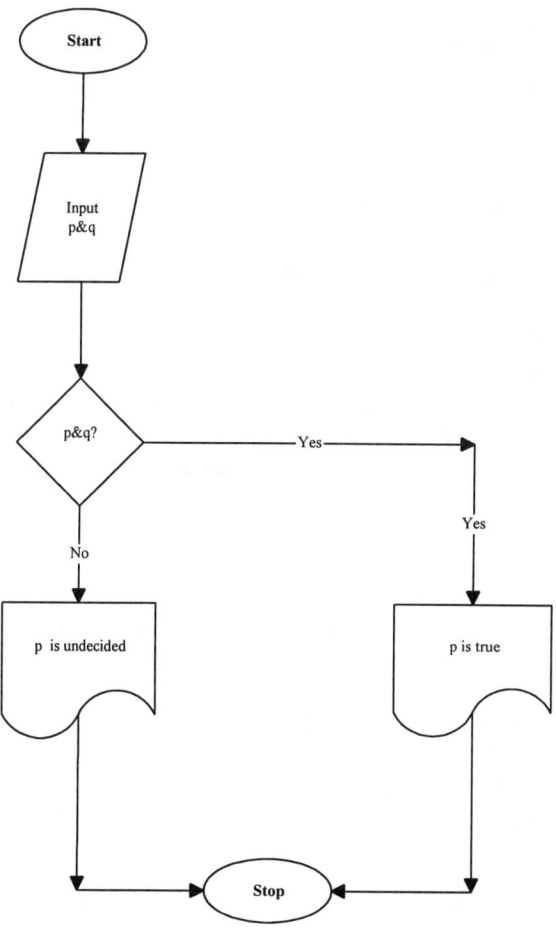

Figure 10.6

Algorithm For Simplification

Given that the value of p & q is known, what can be known about the value of p?

1.   Input value of p & q.
2.   If p & q is false,
         (a) p is undecided.
         (b) Stop.
3.   p is true.
4.   Stop.

Some Additional Algorithms

Below are some slightly more complicated algorithms which make use of the concepts we have been considering. One use of algorithms is as the first step in the development of computer programs. Efficiency here is desirable. Therefore, one goal in the development of an algorithm is to minimize the number of steps required to reach the conclusion which gives the maximum possible information. Understanding the logic of relations is necessary to achieve this efficiency.

Figure 10.7

Algorithm for Determining Value of R, S, T, U given that the value of [(R & S) & (T & U)] is known

1.   Input value of [(R & S) & (T & U)].
2.   If [(R & S) & (T & U)] are true,
         (a) R, S, T, U are true.
         (b) Stop.
3.   R, S, T, U are undecided.
4.   Stop.

Figure 10.8

Given the value of R, S, T, U, what can be known about the value of (R & S), [(R & S) & T] and [(R & S) & (T & U)]

1. Start.
2. Input values of R, S, T, U.
3. If R is false,
    (a) (R & S), [(R & S) &T], [(R & S) & (T & U)] are false.
    (b) Stop.
4. If S is false,
    (a) (R & S), [(R & S) & T], [(R & S) & (T & U)] are false.
    (b) Stop.
5. If T is false,
    (a) (R & S) is true & [(R & S) & T] & [(R & S) & (T & U)] are false.*
    (b) Stop.
6. If U is false,
    (a) (R & S), [(R & S) & T] are true & [(R & S) & (T & U)] are false.
    (b) Stop.
7. (R & S), [(R & S)&T] & [(R & S) & (T & U)] are true.
8. Stop.

The explanation for this algorithm is the mirror image of the previous one. If one part of a string of conjunctions is false then the whole conjunct (although not necessarily all the parts) is false. Thus, if R is false or if S is false, then (R & S), [(R & S) & T], and {(R & S) & (T & U)] must be false; if T is false then [(R & S) & T] and [(R & S) & (T & U)] must be false; etc.

---

* Remember that if a given statement is not false, it is assumed to be true and the algorithm proceeds to the next step. Thus, the algorithm proceeds to step five only because R (step 3) and S (step 4) were found to be true.

Figure 10.9

Algorithm for Determining value of (A →B) → [(R & S) & (T & U)] given the values of A, B, R, S, T and U.

1. Input value of A, B, R, S, T, U
2. If A is true,
        (a) if B is false,
            (A→B) → [(R & S) & (T & U)] is true.
        (b) Stop.
3. If R is false,
        (a) (A→B) → [(R & S) & (T & U)] is false.
        (b) Stop.
4. If S is false,
        (a) (A → B) → [(R & S) & (T & U)] is false.
        (b) Stop.
5. If T is false,
        (a) (A → B) → [(R & S) & (T & U)] is false.
        (b) Stop.
6. If U is false,
        (a) (A→B) → [(R & S) & (T & U)] is false.
        (b) Stop.
7. (A → B) → [(R & S) & (T & U)] is true.
8. Stop.

The algorithm for (A → B) → [(R & S) & (T & U)] is a little more complicated in that we are dealing with both an entailment and a series of conjuncts. We know that if our p statement is false that our entailment is true regardless of the truth or falsehood of the q statement, Thus, we work first with the condition which makes A → B false, namely A being true when B is false. If the p statement (A → B) is true then its entailment can be true only if the q statement [(R & S) & (T & U)] is true. But the q statement can be true only if **all** of the parts of the conjunct are true. This is the reason that steps #3-#6 all indicate (A → B) → [(R & S) & (T & U)] is false.

PROBLEM SET: CONJUNCTION AND SIMPLIFICATION

Prove the validity of the following arguments..

1.
   1. $[(D \rightarrow F) \& (O \& P)] \rightarrow [(R \rightarrow S) \& (Q \rightarrow T)]$ $\therefore (R \rightarrow S)$
   2. $D \rightarrow F$
   3. $O \& P$

2.
   1. $\sim(O \rightarrow P) \rightarrow (H \& W)$ $\therefore H$
   2. $(O \rightarrow P) \rightarrow (R \rightarrow S)$
   3. $\sim(R \rightarrow S)$

3.
   1. $[\sim(O \rightarrow P) \& (T \rightarrow W)] \rightarrow [(H \& W) \rightarrow (R \& S)]$ $\therefore R$
   2. $(T \rightarrow W) \& (A \& B)$
   3. $\sim(R \rightarrow S)$
   4. $(O \rightarrow P) \rightarrow (R \rightarrow S)$
   5. $H \& W$

4.
   1. $\{ \{[P \&(P \rightarrow R)] \& [(T \& W) \& (R \rightarrow S)]\} \& \{[(Z \rightarrow W) \& (S \rightarrow R)] \& [(P \rightarrow R) \rightarrow (R \rightarrow S)]\} \} \rightarrow (A \& B)$ $\therefore A$
   2. P
   3. $(P \rightarrow R)$
   4. T
   5. W
   6. $R \rightarrow S$
   7. $Z \rightarrow W$
   8. $S \rightarrow R$
   9. $T \rightarrow [(P \rightarrow R) \rightarrow (R \rightarrow S)]$

5.
   1. $\{[(S \rightarrow R) \& (P \rightarrow R)] \& (Q \rightarrow R)\} \rightarrow [(R \rightarrow S) \& (R \rightarrow Q)]$
   2. $Q \rightarrow R$ $\therefore R \rightarrow S$
   3. $(S \rightarrow R) \&(P \rightarrow R)$

6.
    1. $[(P \rightarrow R)\&(Q \rightarrow S)] \rightarrow (T \rightarrow W)$ ∴ ~$[(P \rightarrow R) \& (Q \rightarrow S)]$
    2. $(A \rightarrow B) \rightarrow$~$(C \rightarrow D)$
    3. $(T \rightarrow W) \rightarrow (C \rightarrow D)$
    4. $(A \rightarrow B) \& A$

7.
    1. $[$~$(P \rightarrow Q)\&$~$(R \rightarrow S)] \rightarrow ($~$A \&$~$T)$ ∴ ~A
    2. $(P \rightarrow Q) \rightarrow (C \rightarrow D)$
    3.   ~$(C \rightarrow D)$
    4. $(R \rightarrow S) \rightarrow (C \rightarrow D)$

8.
    1. $(A \& B) \rightarrow (C \& D)$ ∴ C
    2. A
    3. $(E \& F) \rightarrow B$
    4. E
    5. F

9.
    1. $(A \& B) \rightarrow (C \& D)$ ∴ ~$(A \& B)$
    2. $(C \& D) \rightarrow (E \& F)$
    3. $(E \& F) \rightarrow (G \& H)$
    4.   ~$(G \& H) \&$~$(I \& J)$

10.
    1. $[(A \& B) \text{ V } (C \& D)] \rightarrow [(H \& G) \& (E \text{ V } F)]$ ∴ H
    2. $[(A \& B) \text{ V } (C \& D)]$

1.   Jones is liable for damages only if Jones is negligent. If Jones is a careful man, then he is neither negligent nor immature. Jones is a careful man. Therefore, Jones is not liable for damages. L, N, C, I.

2.   If Smith worked more than forty hours and if he is not salaried, then he receives overtime pay and he will pay more taxes. Smith worked more than forty hours. Smith is not salaried. Therefore, Smith receives overtime pay. W, S, O, T.

3. If Smith worked more than forty hours and if he is not salaried, then he receives overtime pay and he will pay more taxes. Smith did not both receive overtime pay and pay more taxes. If it is not the case both that Smith worked more than forty hours and that he is not salaried, then he will not be in a higher tax bracket. Therefore, Smith will not be in a higher tax bracket. W, S, O, T, B.

4. If God is not finite, then he is not comprehensible. If God is finite, then he is limited in his power and evil has a rational explanation. If evil has a rational explanation, then Brightman has solved the problem of evil. If evil does not have a rational explanation and God is not limited, then the statement which asserts that God is limited and that evil has a rational explanation is false. Brightman has not solved the problem of evil. Neither has Teilhard. If Teilhard has not solved the problem of evil, then God is not limited in his power. Therefore, God is not comprehensible and evil does not have a rational explanation. F, C, L, R, B, T.

5. If process philosophers are correct, then God is eternal in the sense that he is part of an unending temporal sequence. If Augustine is right, then God is eternal in the sense that he transcends all temporal categories. If God is eternal in the sense that he is part of an unending temporal sequence, then he is subject to time as we are. If God is subject to time as we are, then he is not all knowing and he is not omnipotent. It is false to say that God is not all knowing and not omnipotent. If process philosophers are not correct, then Augustine is correct. If God is eternal in the sense that he transcends all temporal categories, then God knows in a way different from us and he sees the past, present and future as an eternal now. If God knows in a way different from the way we know, then he is incomprehensible. Therefore, God is incomprehensible.
P, E1, A, E2, S, K, O, D, N, I.

6. If the patient has symptom #1 and symptom #2, then the patient has disease #1 and symptom #3. If the patient has disease #1 and symptom #3, then his symptom may be treated by medicine #1 and his disease may be cured by medicine #2. The patient has symptom #2. The patient has symptom #1. Therefore, the patient may be cured by medicine #1.
S1, S2, D1, S3, M1, M2.

7. If the red book is between the yellow and the blue books, then the red book is not last. If there are only four books, then the fourth book is last. There are only four books. If the red book is not last and if the fourth book is last, then the red book is not fourth. The red book is between the yellow and the blue books. Therefore, the red book is not fourth.

    B, R, F, L, A.

8. If the blue book is between the orange and the red book, then the orange and blue books are next to each other. The blue book is between the orange and the red book. If the red book is between the yellow and the blue book, then the red and yellow books are next to each other. The red book is between the yellow and the blue book. Therefore, the orange and blue books are next to each other, and the red and yellow books are next to each other. B, O, R, N.

9. If Jones' legal liability means Smith will get his money and if Jones is legally liable, then Z Corporation will suffer financial loss. If Z Corporation suffers financial loss, Z Corporation's stockholders will lose money and Z Corporation's stock will go down. If Z Corporation's stockholders lose money, then Brown will go bankrupt. Smith will get his money if Jones is legally liable. Jones is legally liable. Therefore, Brown will go bankrupt. J, S, F, L, D, B.

10. Rule one (anti discrimination law) is relevant only if the Mall Owners Association wishes to exclude union pickets on grounds that the union has a disproportionate number of minorities. Rule two (regulation of union activity law) is relevant if the union pickets have been obstructive, and rule two is relevant if the union pickets have been threatening. The union pickets have been obstructive but not threatening. The Mall Owners Association wishes to exclude union pickets on the grounds that the union has a disproportion number of minorities only if the Mall Owners Association is all white, Anglo-Saxon Protestants. The Mall Owners Association is not all white, Anglo-Saxon Protestants. Therefore, rule one is not relevant, and rule two is relevant.

    R1, M, R2, O, T, W.

CHAPTER 11: HYPOTHETICAL SYLLOGISM

The principle of hypothetical syllogism which is abbreviated "H.S." is depicted below.

Figure 11.1

p	→	q
q	→	r
∴ p	→	r

If I have a car, then I have transportation. If I have transportation, then I have a way to get to the game. Therefore, if I have a car, then I have a way to get to the game. C, T, G.

Figure 11.2

1	C	→	T
2	T	→	G
∴	C	→	G

Below are depicted a truth table, a flow chart and an algorithm for hypothetical syllogism or H.S.

Figure 11.3

Truth Table For H.S.

p	q	r	[(p	→	q)	&	(q	→	r)]	→	(p	→	r)
1	1	1	1	1	1	1	1	1	1	1	1	1	1
1	1	0	1	1	1	0	1	0	0	1	1	0	0
1	0	1	1	0	0	0	0	1	1	1	1	1	1
1	0	0	1	0	0	0	0	1	0	1	1	0	0
0	1	1	0	1	1	1	1	1	1	1	0	1	1
0	1	0	0	1	1	0	1	0	0	1	0	1	0
0	0	1	0	1	0	1	0	1	1	1	0	1	1
0	0	0	0	1	0	1	0	1	0	1	0	1	0

Figure 11.4

Flow Chart For Hypothetical Syllogism

Given the value of p → q and q → r, what can be known about p → r?

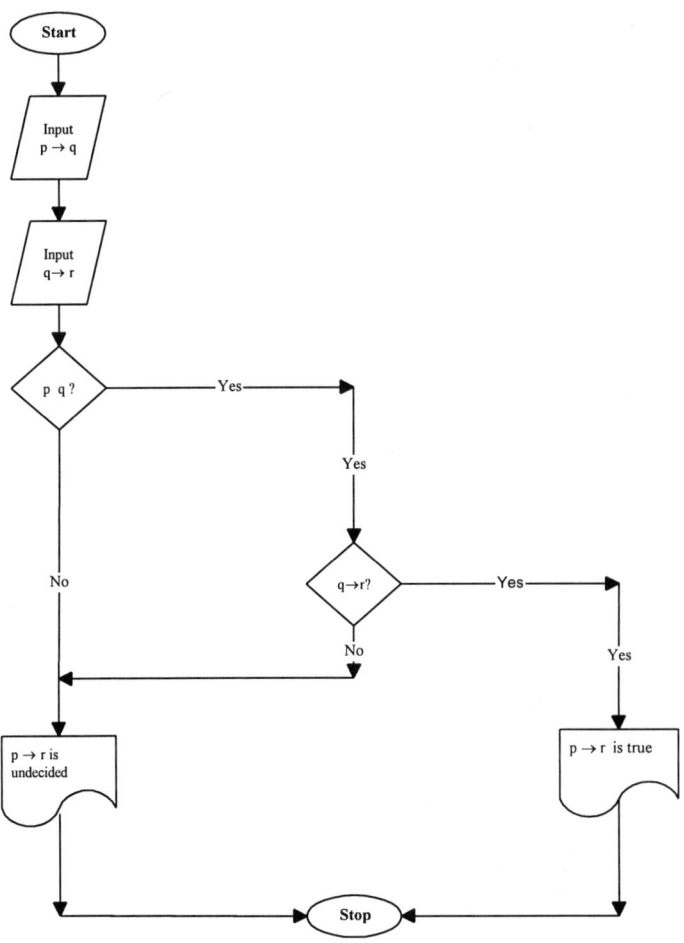

Figure 11.5

Algorithm For Hypothetical Syllogism
Given the value $p \to q$ and $q \to r$, what can be known about $p \to r$?

1.   Input value of $p \to q$.
2.   Input value of $q \to r$.
3.   If $p \to q$ is false,
        (a) value of $p \to r$ is undecided.
        (b) Stop.
4.   If $q \to r$ is false,
        (a) value of $p \to r$ is undecided.
        (b) Stop.
5. $p \to r$ is true.
6.   Stop.

PROBLEM SET: HYPOTHETICAL SYLLOGISM
    Demonstrate the validity of the following arguments. Use H.S. whenever possible.

1.
    1. $D \to C$ $\therefore D \to Y$
    2. $C \to R$
    3. $R \to S$
    4. $S \to T$
    5. $T \to U$
    6. $U \to V$
    7. $V \to X$
    8. $X \to Y$

2.
    1. $(A \to B) \to (C \to D)$ $\therefore (A \to B) \to (E \to F)$
    2. $(C \to D) \to (E \to F)$

3.
1. $[(A \rightarrow B) \& (C \rightarrow D)] \rightarrow [(E \rightarrow F) \& (G \rightarrow H)]$

2. $[(E \rightarrow F) \& (G \rightarrow H)] \rightarrow [(I \rightarrow J) \& (K \rightarrow L)]$
3. $\sim [(I \rightarrow J) \& (K \rightarrow L)]$

$\therefore \sim [(A \rightarrow B) \& (C \rightarrow D)]$

4.
1. $\{[(A \rightarrow B) \& (C \rightarrow D)] \rightarrow [(E \rightarrow F) \rightarrow (I \rightarrow J)]\} \rightarrow K$
2. $K \rightarrow L$

$\therefore \{[(A \rightarrow B) \& (C \rightarrow D)] \rightarrow [(E \rightarrow F) \rightarrow (I \rightarrow J)]\} \rightarrow L$

5.
1. $\{[(A \rightarrow B) \& (C \rightarrow D)] \rightarrow [(E \rightarrow F) \rightarrow (I \rightarrow J)]\} \rightarrow$
   $\{[(K \rightarrow L) \& (M \rightarrow N)] \rightarrow (O \rightarrow P)\}$
2. $Q \rightarrow \{[(A \rightarrow B) \& (C \rightarrow D)] \rightarrow [(E \rightarrow F) \rightarrow (I \rightarrow J)]\}$
3. $\{[(K \rightarrow L) \& (M \rightarrow N)] \rightarrow (O \rightarrow P)\} \rightarrow R$
4. $R \rightarrow S$　　　　　　　　　$\therefore Q \rightarrow S$

6.
1. $F \rightarrow G$　　　　　　　　　$\therefore A \rightarrow I$
2. $C \rightarrow D$
3. $E \rightarrow F$
4. $G \rightarrow H$
5. $B \rightarrow C$
6. $H \rightarrow I$
7. $A \rightarrow B$
8. $D \rightarrow E$

7.
1. $[(H \& I) \& (J \& K)] \rightarrow \{[(L \& M) \& (N \& O)] \& (P \& Q)\}$
2. $[(E \& F) \& G] \rightarrow [(H \& I) \& (J \& K)]$
3. $(C \& D) \rightarrow [(E \& F) \& G]$
4. $\{[(L \& M) \& (N \& O)] \& (P \& Q)\} \rightarrow$
   $\{[(R \rightarrow S) \rightarrow (T \rightarrow U)] \& (V \rightarrow W)\}$
5. $B \rightarrow (C \& D)$
6. $\{[(R \rightarrow S) \rightarrow (T \rightarrow U)] \& (V \rightarrow W)\} \rightarrow (X \rightarrow Y)$
7. $A \rightarrow B$
8. $(X \rightarrow Y) \rightarrow Z$　　　　　　　$\therefore A \rightarrow Z$

8.

    1.   A → { {[(B → C) → (D → E)] → [(F → G) → (H → I)]} →
                  {[(J → K) → (L → M)] → [(N → O) → (P → Q)]} }

    2.   { {[(Z → Y) → (X → W)] → [(V → W) → (T → S)]} →
               {[R → Q) → (P → O)] → [(N → M) → (L → K)]} } → A

    ∴   { {[(Z → Y) → (X → W)] → [(V → W) → (T → S)]} → {[R → Q)
             → (P → O)] → [(N → M) → (L → K)]} } →
        { {[(B → C) → (D → E)] → [(F → G) → (H → I)]} →
        {[(J → K) → (L → M)] → [(N → O) → (P → Q)]} }

9.

    1.   (C & D) → (E & F)   ∴ H
    2.   (A & B) → (C & D)
    3.   (E & F) → (H & I)
    4.   B
    5.   A

10.

    1.   G   ∴ ~(A & B)
    2.   H
    3.   (C & D) → (E & F)
    4.   (A & B) → (C & D)
    5.   (G & H) → ~(E & F)

1.   If God is perfect, then he is omniscient. If God is omniscient, then he can predict the future. If God can predict the future, then the future is knowable. If the future is knowable, then man is determined. If man is determined, then he is not morally responsible. If man is not morally responsible, then he has not sinned. If man has not sinned, then man is perfect. Therefore, if God is perfect, then man is perfect.

                      G, O, F, K, D, M, S, M1 .

2.   If Jones works overtime, then he is classified as an hourly employee. If Jones is an hourly employee, then he must be compensated at time and 1/2 for every hour over forty. If Jones is compensated at time and one half for every hour over forty, then his income will place him in a higher tax withholding schedule. Therefore, if Jones works overtime, he will be in a higher tax withholding schedule.   O, H, C, T.

3. If Jones works 20 hours overtime, then his weekly income will be over $500. If his weekly income is over $500, then he will be in deduction schedule D3. If he is in deduction schedule D3, then he will have more withheld on an annual basis than is necessary. If Jones is having an excessive amount withheld, then he should claim another deduction. Jones should not claim another deduction. Therefore, Jones does not work 20 hours overtime. O, I, D3, M, A.

4. Jones would not be guilty of driving without a license unless he were drunk. Jones would be drunk only if the bartender had continued to serve Jones after he had had too much and Smith had persuaded him to go to the bar. Therefore, if Jones is guilty of the crime of driving without a license, then the bartender continued to serve him after he had had too much and Smith persuaded him to go to the bar. G, D, B, P,

5. Jones being drunk is a necessary condition of his being guilty of driving without a license. Jones being drunk is a sufficient condition for the bartender having continued to serve him after he had had too much and for Smith to have persuaded him to go to the bar. The bartender is legally liable if Jones were in an accident and if the bartender had continued to serve him after he had had too much. Jones was guilty of driving without a license. Jones was in an accident. Therefore, the bartender is legally liable. D, G, B, P, L, A.

6. If Jones got drunk, then he had an accident. If Smith persuaded Jones to go to the bar, then the bartender continued to serve Jones after he had had too much. If Jones had an accident, then the bartender is legally liable. If the bartender continued to serve Jones after he had had too much, then Jones got drunk. Therefore, if Smith persuaded Jones to go to the bar, the bartender is legally liable. D, A, P, B, L.

7. If the bartender continued to serve Jones after he had had too much and if Jones got drunk, then if Jones has an accident then the bartender is legally liable. If Smith persuaded Jones to go to the bar, then the bartender continued to serve Jones after he had had too much and Jones got drunk. If Jones having an accident entails that the bartender is legally liable, then the bartender's license may be revoked. Therefore, if Smith persuaded Jones to go to the bar, the bartender's license may be revoked. B, D, A, L, P, R.

8. If the patient has disease #1, then the patient has symptom #1. If the patient has symptom #1, then the patient has symptom #2. If the patient has symptom #2, then the patient has symptom #3. The patient does not have symptom #3. Therefore, the patient does not have disease #1.

D1, S, S2, S3.

9. The patient has disease #1 only if the patient has symptom #1. The patient's having symptom #1 is a sufficient condition for the patient to have symptom #2. The patient's having symptom #3 is a necessary condition for the patient's having symptom #2. It is not the case that the patient has symptom #3. Therefore, the patient does not have disease #1.

D1, S1, S2, S3.

10. Paraphrase of I Corinthians 15: 12-20.

If there is no resurrection from the dead, then Christ has not been raised. If Christ has not been raised then our preaching is useless. If our preaching is useless then your faith is futile. If your faith is futile then you are yet in your sins. If you are yet in your sins, then those who have fallen asleep in Christ are lost. If those who have fallen asleep in Christ are lost, then we have hope only in this life. If we have hope only in this life, then we are to be pitied more than all men. We are not to be pitied more than all men. Therefore, it is not the case that there is no resurrection from the dead.   R, C, U, F, S, A, H, P.

CHAPTER 12: ABSORPTION

The principle of absorption is abbreviated "Abs." and is depicted below.

Figure 12.6

	p	→	q
∴	p	→	(p & q)

Figure 12.2

If I have a cat, then I have a dog. Therefore, if I have a cat, then I have a cat and a dog. C, D.

1.	C	→	D
∴	C	→	(C & D)

Below are depicted a truth table, a flow chart and an algorithm for absorption (Abs).

Logic of the Computing Sciences

Figure 12.3

Truth Tables For Absorption

p	q	(p	→	q)	→	[p	→	(p	&	q)]
1	1	1	1	1	1	1	1	1	1	1
1	0	1	0	0	1	1	0	1	0	0
0	1	0	1	1	1	0	1	0	0	1
0	0	0	1	0	1	0	1	0	0	0

As truth tables and our flow chart and algorithm indicate,
(p → q) → [p → (p & q)] is true regardless of whether p → q is true or
false. This is true because it is impossible to have a situation in which
p → q is true simultaneously with p → (p & q) being false. When p →
q is false, however, p → (p & q) is also false. This is shown by truth
tables as well as in the flow chart and algorithm.

Figure 12.4

Flow Chart For Absorption

Given the value of p → q, what can be know about the value of
p → (p & q)?

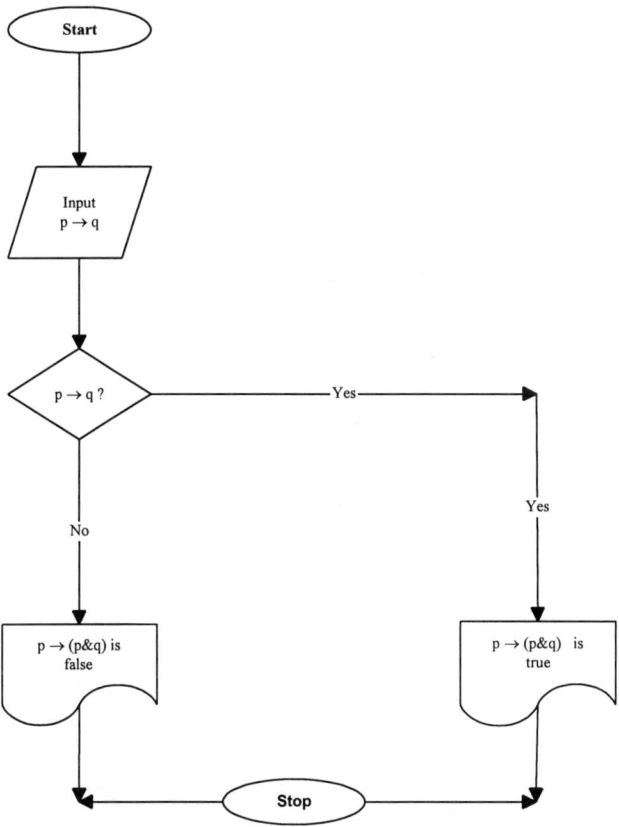

Figure 12.5

Algorithm for Absorption

Given the value of p → q, what can be known about the value of
p → (p & q)?

  1.   Input value of p → q.
  2.   If p → q if false,
           (a) p → (p & q) is false.
           (b) Stop.
  3.   p → (p & q) is true.
  4.   Stop.

That if p → q is false, then p → (p & q) must also be false can be
shown by truth tables. For p → q to be false our p statement must be
true while our q statement is false. This will result in p → (p & q)
being false. This is shown below in figure 12.6.

Figure 12.6

p	q	p	→	q
1	0	1	0	0

p	→	(p	&	q)
1	0	1	0	0

PROBLEM SET: ABSORPTION
Demonstrate the validity of the arguments given below. (Use Absorption in
solution when possible).

1. 1.   D → C   ∴ D → (D & Y)
    2.   C → R
    3.   R → S
    4.   S → T
    5.   T → U
    6.   U → V
    7.   V → X
    8.   X → Y

2. 1. D → C   ∴ D → [(S & T) & U]
    2. C → R
    3. (C & R) → S
    4. S → T
    5. (S & T) → U

3. 1. D → C   ∴~D
    2. C → R
    3. (C & R) → S
    4. S → T
    5. (S & T) → U
    6. ~ [(S & T) & U]

4. 1. D → C   ∴ U
    2. C → R
    3. (C & R) → S
    4. S → T
    5. (S & T) → U
    6. D

5. 1. D → C   ∴ ~C
    2. C → R
    3. (C & R) → S
    4. S → T
    5. (S & T) → U
    6. ~ [(S & T) & U]

6. 1. ~M →~ P   ∴ ~ M → (~ M &~ P)

7. 1. C → D   ∴ C → C & {D & [E & (F & G)]}
    2. D → E
    3. E → F
    4. F → G

8. 1. D → E   ∴ A → A & {B & [C & (D & E)]}
    2. A → B
    3. C → D
    4. B → C

9. 1. E → F   ∴ F & [(G → H) & I]
    2. F → (G → H)
    3. (G → H) → I
    4. E   (Solve without using absorption.)

10. 1.   E → F   ∴ F & [(G → H) & I]
     2.   F → (G → H)
     3.   (G → H) → I
     4.   E   (Solve without using Conjunction.)

1.   If Jones works overtime, then he is classified as an hourly employee. If Jones is an hourly employee, then he must be compensated at time and 1/2 for every hour over forty. If Jones is compensated at time and one half for every hour over forty, then his income will place him in a higher tax withholding schedule. Therefore, if Jones works overtime, then he is classified as an hourly employee, and he must be compensated at time and one half for every hour over forty and he will be in a higher tax withholding schedule. O, H, C, T.

2.   Jones' not having a valid driver's license is a necessary condition for his being guilty of the crime of driving without a license. Jones would not be guilty of driving without a valid driver's license unless he were drunk. Jones would be drunk only if Smith had persuaded him to go to the bar and the bartender had continued to serve Jones after he had had too much. If Jones were in an accident and if the bartender had continued to serve him after he had had too much, then the bartender is liable under the law for Jones' accident. Smith's having persuaded Jones to go to the bar is not a sufficient condition for making Smith liable under the law for Jones' accident. It is a sufficient condition for making him morally liable. Jones was guilty of the crime of driving without a license. Jones was in an accident. If the bartender was liable under the law, then his license to sell liquor may be revoked. Therefore, if Jones had an accident and the bartender continued to serve him after he had had too much, then the bartender is liable under the law and his license to sell liquor may be revoked. V, G, D, P, B, A, L, S, M, R.

3.   If the patient has disease #1, then the patient has symptom #1. If the patient has symptom #1, then the patient has symptom #2. If the patient has symptom #2, then the patient has symptom #3. The following is not the case: the patient has symptom #1, and symptoms #2 and #3. Therefore, the patient does not have disease #1. D1, S1, S2, S3.

4. The patient has disease #1 only if the patient has symptom #1. The patient's having symptom #1 is a sufficient condition for the patient to have symptom #2. The patient's having symptom #3 is a necessary condition for the patient's having symptom #2. The following is not the case: the patient has symptom #1, and #2 and #3. Therefore, the patient does not have disease #1. D1, S1, S2, S3.

5. Jones is on salary schedule #1, only if he is an hourly employee. Jones is not an hourly employee, unless he is classified as unskilled. If Jones is classified as an hourly and an unskilled employee then he has to be in a training program. Jones's being in a training program is a sufficient condition for his being placed on third shift. If Jones is on third shift, then a 10% increment is added to his regular salary. Therefore, if Jones is on salary schedule #1, he is on third shift and receives a 10% increment. S, H, U, P, T, I.

6. Jones is an hourly employee, if he is on salary schedule #1. Jones is classified as an hourly employee and as unskilled employee only if he is in a training program. If Jones is classified as an hourly employee, then he is classified as unskilled. Jones's being on third shift is a necessary condition for his being in a training program. Jones's being on third shift is a sufficient condition for a 10% increment being added to his salary. Jones is not both in a training program and having a 10% increment added to his salary. Therefore, Jones is not on salary schedule #1.
    H, S, U, P, T, I.

7. Plato's view of mathematics is correct only if mathematics is descriptive of the universe. Plato's view of mathematics is correct and mathematics is descriptive of the universe only if it is not the case that Ayer's view of mathematics is correct. Plato's view of mathematics is correct. If there is no such thing as synthetic a priori propositions, then Ayer's view of mathematics is correct. Therefore, it is not the case that there is no such thing as synthetic a priori propositions. P, D, A, S.

8. If Jones is a cultural relativist, then he believes that all values are dependent upon culture. Jones believes that all values are dependent on culture only if he believes that no values transcend the culture and believes there are no absolute values. Therefore, Jones's being a cultural relativist is a sufficient condition for the following: his believing that all values are dependent on culture, and believing no values transcend the culture and believing there are no absolute values. R, D, T, A.

9. If Jones is a cultural relativist, then he believes that there are no values which all cultures have in common. If Jones believes that there are no values which all cultures have in common, then he also believes that the fact that there are no values which all cultures have in common constitutes evidence for the fact that there are no transcendent values. Therefore, if Jones is a cultural relativist, then he believes that there are no values which all cultures have in common and the fact that there are no values which all cultures have in common constitutes evidence for the fact that there are no transcendent values.

    Let **R** equal Jones is a cultural relativist.

    Let **C** equal Jones believes that there are no values which all cultures have in common

    Let **T** equal the <u>fact</u> that there are no values which all cultures have in common.

    Let **E** equal Jones believes **T** constitutes evidence for the fact that there are no transcendent values.

10. If Jones is a cultural relativist, then he believes that there are no absolute values and believes that the denial of absolute values leads to tolerance. If Jones believes that there are no absolute values and that the denial of absolute values leads to tolerance, then he believes that tolerance is a good. Jones believes that tolerance is a good only if he believes that it is a good relative to a culture. Therefore, Jones is a cultural relativist only if he believes that there are no absolute values and that the denial of absolutes values leads to tolerance, and that tolerance is a good and that it is a good relative to culture.   R, A, D, T, G.

CHAPTER 13: THE INCLUSIVE OR

In the following three chapters we are going to consider three new rules of inference. These are Disjunctive Syllogism which is abbreviated D.S., Constructive Dilemma which is abbreviated C.D. and Addition which is abbreviated Add. These all require the understanding of a new symbol "V" which symbolizes the English word "or". More specifically this is the symbol for what is known as the "inclusive disjunction."

Although you may not be familiar with the terms "inclusive" and "exclusive disjunction" you doubtless are familiar with the concept. For example, if you are invited out to dinner and your host or hostess asks if you want coffee or tea, what is intended is doubtless an exclusive disjunction. That is to say, an affirmative answer implies either that you want tea **and not coffee** or that you want coffee **and not tea.** The choice of one **excludes** the choice of the other. On the other hand, a follow up question "Do you take cream or sugar?" usually is intended as an inclusive disjunction. An affirmative answer leaves open three possibilities. You may take cream and not sugar. You may take sugar and not cream. You may take both cream and sugar. We will use the symbol "**V**" to indicate the inclusive disjunction. Thus, if we say of Jones that he has a cat or a dog, it would be symbolized as C **V** D and would affirm that Jones **at least** has one or the other. This condition would be met under the following conditions:

(1) Jones has a cat and a dog.
(2) Jones has a cat but does not have a dog.
(3) Jones does not have a cat but does have a dog.

C V D would not choose between conditions (1), (2) and (3). It would only affirm that at least one of them is true. The inclusive disjunction is illustrated below by a truth table, a flow chart and an algorithm.

Figure13.1

Truth Table For The Inclusive OR

p	q	p	V	q
1	1	1	1	1
1	0	1	1	0
0	1	0	1	1
0	0	0	0	0

As the truth table shows, the <u>only</u> condition under which p v q is false is when **both** p and q are false.

Below are given flow charts and algorithms for the "inclusive or." Explain their structure by using truth tables and ordinary English.

Figure 13.2

Flow Chart For Inclusive "OR" #1

Given the values of p and of q, what can be known about the value of p v q?

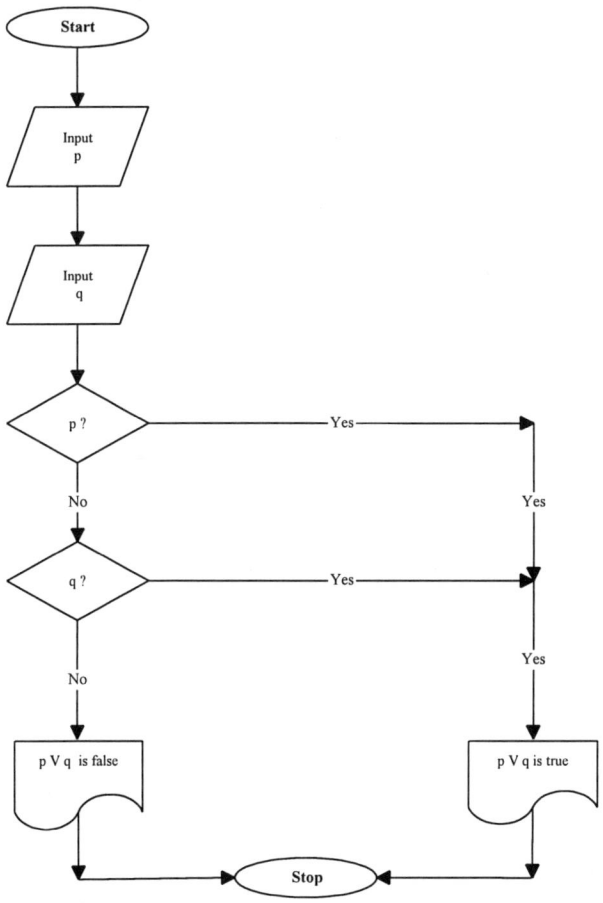

Figure 13.3

Algorithm For Inclusive OR #1

Given the values of p and of q, what can be known about the value of
p v q?

1.   Input the value of p.
2.   Input the value of q.
3   If p is true,
         (a) p V q is true.
         (b) Stop.
4.   If q is true,
         (a) p V q is true.
         (b) Stop.
5.   p V q is false.
6.   Stop.

Figure 13.4

Flow Chart For Inclusive OR #2

Given the values of C, D, E and F, what can be known about the value of
(C V D) V (E V F)?

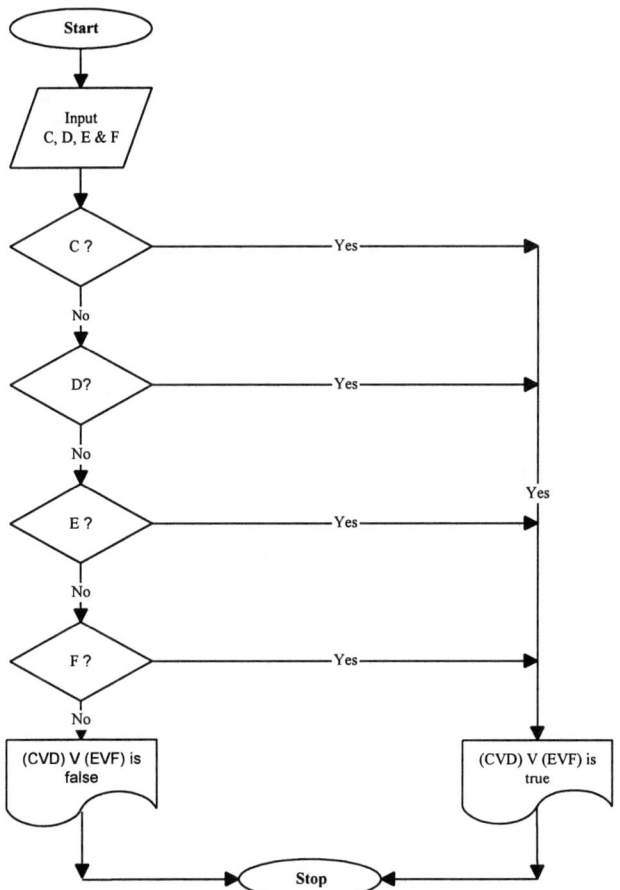

Figure 13.5

Algorithm For Inclusive OR #2

Given the values of C, D, E, and F, what can be known about the value of
(C V D) V (E V F)

1.   Input the value of C, D, E and F.
2.   If C is true,
              (a) (C V D) V (E V F) is true.
              (b) Stop.
3   If D is true,
               (a) (C V D) V (E V F) is true.
              (b) Stop.
4.   If E is true,
               (a) (C V D) V (E V F) is true.
              (b) Stop.
5.   If F is true,
               (a) (C V D) V (E V F) is true.
              (b) Stop.
6.   (C V D) V (E V F) is false.
7.   Stop.

Figure 13.6

Flow Chart for Conjunction And Inclusive OR #3

Given the values of C, D, E and F, what can be known about the value of
(C & D) V (E & F)?

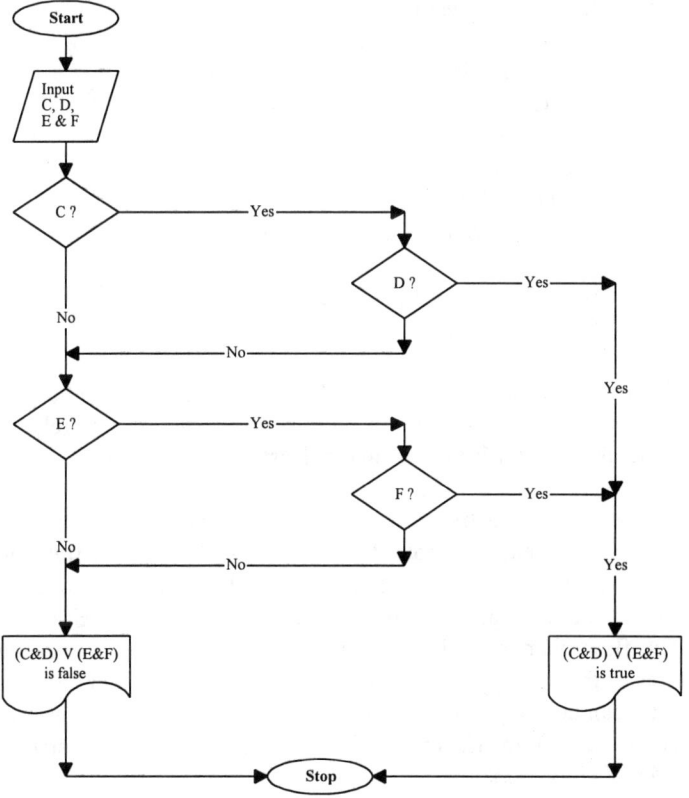

Figure 13.7

Algorithm For Conjunction And Inclusive OR #3

Given the values of C, D, E and F, what can be known about the
value of (C & D) V (E & F)?

1. Input the value of C, D, E and F.
2. If C is true,
       Then if D is true,
            (a) (C & D) V (E & F) is true.
            (b) Stop.
3. If E is true,
       Then if F is true,
            (a) (C & D) V (E & F) is true.
            (b) Stop.
4. (C & D) V (E & F) is false.
5. Stop.

The flow chart and algorithm for (C V D) V (E V F) can be readily
understood once the principle of the "inclusive or" is comprehended.
The "inclusive or" affirms only that at least one of statements indicated
is true.

If for example C is true, then the whole statement C V D) is true. If
(C V D) is true then, the whole statement [(C V D) V (E V F)] is true.

It can readily be seen that the same principle works if any one of the
statements indicated are true. It is only false if they are all false.

The flow chart and algorithm for (C & D) V (E & F) combines
features of the principle of conjunction and the "inclusive or".

(C & D) can be true only if assertion C and assertion D are true. But if
either (C & D) or (E & F) are true, then the whole statement
[(C& D) V (E & F)] is true.

CHAPTER 14: DISJUNCTIVE SYLLOGISM

Disjunctive syllogism, abbreviated D.S., like all of the rules of inference we have studied is intuitively obvious. It is symbolized in 14.1 below. 14.2 symbolizes a particular argument which takes the form of disjunctive syllogism.

Figure 14.1

p	V	q
~p		
∴ q		

Figure 14.2

Jones has a cat or a dog.
He doesn't have a cat.
Therefore, he has a dog.

C	V	D
~C		
∴ D		

As indicated in figure 14.3 below, C V D affirms that at least one of three possibilities is true. (1) Jones has a cat and a dog. (2) Jones has a cat and does not have a dog. (3) Jones does not have a cat but does have a dog. Thus, C V D excludes only logical possibility number 4. When we add ~C, this excludes possibility (1) and possibility (2). Therefore, logical possibility number 3, in which D is true, necessarily follows.

Figure14.3

	C	D	C V D
1.	1	1	1
2.	1	0	1
3.	0	1	1
4	0	0	0

As in previous cases we will illustrate this by a truth table, a flow chart and an algorithm.

Figure 14.4

Truth Table For Disjunctive Syllogism

p	q	[(p	V	q)	&	~p]	→	q
1	1		1		0	0	1	1
1	0		1		0	0	1	0
0	1		1		1	1	1	1
0	0		0		0	1	1	0

Figure 14.5

Flow Chart For Disjunctive Syllogism

Below are given a flow chart and algorithm for disjunctive syllogism. Explain their structure using truth tables and ordinary English.

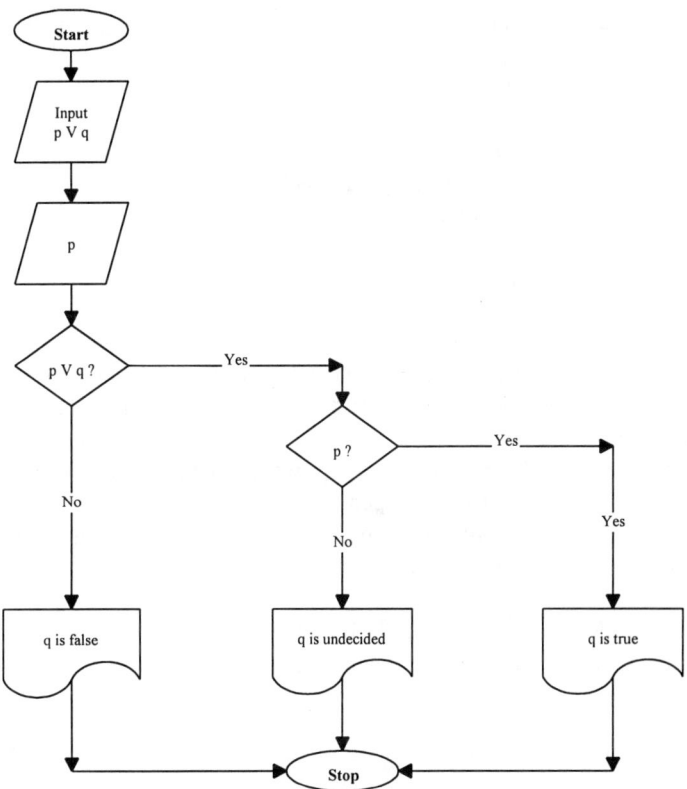

Figure 14.6

Algorithm For Disjunctive Syllogism

Given the value of p V q and p, what can be known about q?

1.   Input value of p V q.
2.   Input value of p.
3.   If p V q is false,
         (a) value of q is false.
         (b) Stop.
4.   If p is true,
         (a) value of q is undecided.
         (b) Stop.
5.   q is true.
6.   Stop.

The flow chart and algorithm for D.S. can be explained as follows. p V q can be false, only if both p and q are false. Therefore, a false p V q indicates both a false p **and** a false q. Therefore a false p V q indicates a false q. A true p V q indicates that at least either p or q is true. If we couple this with a false p, we know that q must be true. If, however, p V q is true, it is possible that both p and q are true. If, therefore, p is true q may be either true or false. Its value is undecided based on the information we have at our disposal.

PROBLEM SET: DISJUNCTIVE SYLLOGISM
Demonstrate the validity of the arguments given below.

1.
     1.   B V A   ∴ A
     2.   ~B

2.
     1.   ~B   ∴ A & C
     2.   B V (A & C)

3.
  1. ~F   ∴ A
  2. F V [(A & B) & C]

4.
  1. G → H   ∴ D
  2. G V [(A → D) & F]
  3. A
  4. ~H

5.
  1. G → H   ∴ ~ A
  2. G V [(A → D) & F]
  3. ~(G & H)
  4. ~D

6.
  1. [(G → H) & (I → J)] V [C V (B → A)] ∴ A
  2. L → M
  3. [R → (L & M)] → ~[(G → H) & (I → J)]
  4. ~C
  5. B
  6. R → L

7.
  1. A V [(C & D) & (R → S)]   ∴ C
  2. (E & F) → (~ G & H)
  3. F
  4. E
  5. A → G

8.
  1. R   ∴ I
  2. [(A V B) & (C V D)] → {(E V F) V [G V (H V I)]}
  3. ~ K
  4. A V B
  5. J → ~ (E V F)
  6. C V D
  7. J
  8. R → (~ H & T)
  9. G → K

9.
      1. ~ (C V D)   ∴ A
      2. ~ (G V H)
      3. I V {(G V H) V [(C V D) V (B V A)]}
      4. ~ I
      5. ~ B

10.
      1. ~ G   ∴ H
      2. [D → (E V F)] V [H & (I → J)]
      3. [D → (E V F)] → G

1.   Jones was either driving too fast or driving on the wrong side of the road. He was not driving to fast. Therefore, Jones was driving on the wrong side of the road. F, W.

2.   If Jones's blood alcohol content was over the legal limit, then he will either go to jail or pay a fine. He will not go to jail. Jones's blood alcohol content was over the legal limit. Therefore, Jones will pay a fine.
                          L, J, F.

3.   Either conviction for driving while legally drunk is a sufficient condition for loosing one's license to drive, or the law is too lenient. It is not the case that conviction for driving while legally drunk is a sufficient condition for loosing one's license. Therefore, the law is too lenient.
                          D, L, A.

4.   Either conviction for driving while legally drunk is a sufficient condition for losing one's license to drive, or paying a fine is a necessary condition for conviction for driving while legally drunk. Conviction for driving while legally drunk is not a sufficient condition for losing one's license to drive. Therefore, one can be convicted for driving while legally drunk only if one pays a fine.   D, L, F.

5.   Either believing that pleasure is a good is a sufficient condition for being a hedonist, or one is a hedonist only if he believes that pleasure is **the** good. Believing that pleasure is **a** good is not a sufficient condition for being a hedonist. Therefore, believing that pleasure is **the** good is a necessary condition for one to be a hedonist.   A, H, T.

6. If Bentham believed that all men always **do** seek pleasure, then Bentham was a psychological hedonist. If Bentham believed that all men always **ought** to seek pleasure then Bentham was an ethical hedonist. Bentham was either not a hedonist of any type, or he believed that all men do seek pleasure. If Bentham believed that all men do seek pleasure then he believed that all men ought to seek pleasure. It is false to say that Bentham was not a hedonist of any type. Therefore, Bentham was both a psychological and an ethical hedonist.   D, P, O, E, H.

7. Bentham was either a qualitative hedonist or a quantitative hedonist. If he was a qualitative hedonist, then he believed that some pleasures are intrinsically better than other pleasures. If he was a quantitative hedonist, then he believed that it was the amount of pleasure which is the only relevant ethical difference between various actions. Bentham did not believe that some pleasures are intrinsically better than others. Therefore, Bentham was a quantitative hedonist and he believed that it was the amount of pleasure which is the only relevant ethical difference between various actions.   B, C, I, A.

8. If Rawls is right, then equality is an intrinsic value. If equality is an intrinsic value, then pleasure is not the only intrinsic value. If pleasure is not the only intrinsic value, then Bentham is wrong. Either the utilitarians are right or Rawls is right. The utilitarians are not right. Therefore, equality is an intrinsic value, and pleasure is not the only intrinsic value and Bentham is wrong.   R, I, P, B, U.

9. All men are not motivated by the love of pleasure. If Nietzsche is wrong, then all men are not motivated by the will to power. All men are rational; or they are motivated by the love of pleasure, or they are motivated by the will to power or ought to be motivated by the love of God. All men are not rational. Nietzsche is wrong. Therefore, all men ought to be motivated by love of God.   P, N, W, R, L.

10. The desired information is either in file one or it is in the following: file two or file three; or file four or five. The desired information is not in file four. The desired information is not in file two or file three. The desired information is not in file one. Therefore, the desired information is in file five.   F1, F2, F3, F4, F5,.

CHAPTER 15: CONSTRUCTIVE DILEMMA

Constructive Dilemma, abbreviated C.D. is symbolized as follows:

Figure 15.1

(p	→	q)	&	(r	→	s)
p	V	r				
∴ q	V	s				

Figure 15.2

If I go to the student center then I will have a cup of coffee. If I go home, then I will have a cup of tea. I am either going to the center or I am going home. Therefore, I am either going to have a cup of coffee or a cup of tea. S, C, H, T.

(S	→	C)	&	(H	→	T)
S	V	H				
∴ C	V	T				

Constructive Dilemma or C.D. makes three affirmations from which the conclusion follows. If p then necessarily q. If r then necessarily s. Necessarily p or r. Therefore, q or s must be true. This principle is depicted below in the form of a flow chart and an algorithm.

Figure 15.3

Flow Chart For Constructive Dilemma
Given the value of p → q, r → s and p V r, what can be known about
q V s?

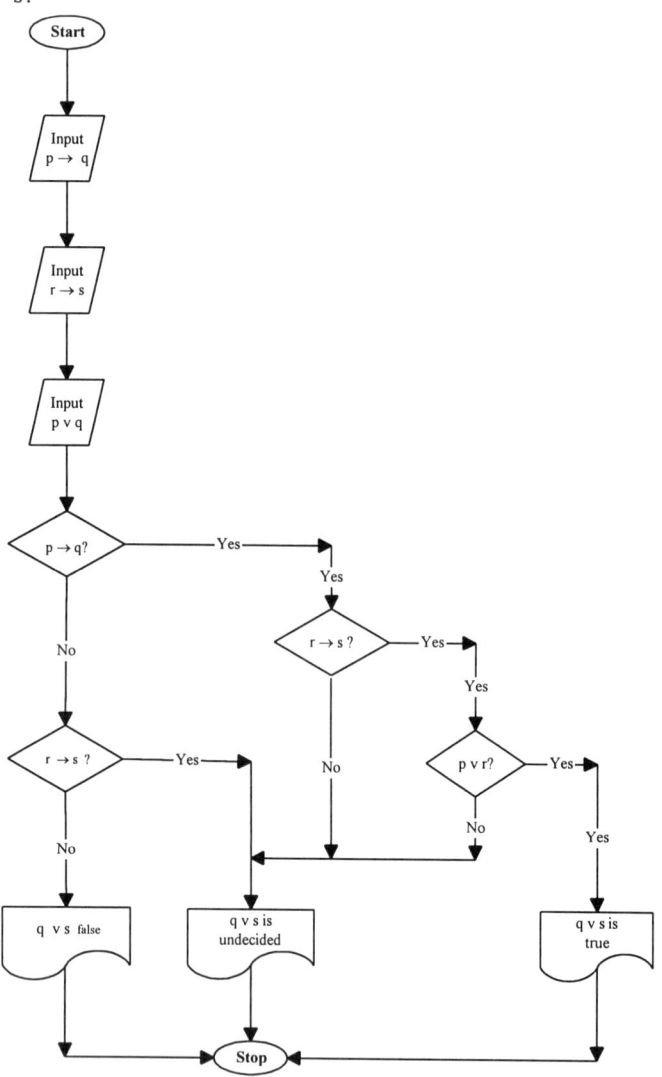

Figure 15.4

Algorithm For Constructive Dilemma
Given the value of $p \to q$, $r \to s$ and $p \lor r$, what can be known about $q \lor s$?

1. Input value of $p \to q$.
2. Input value of $r \to s$.
3. Input value of $p \lor r$.
4. If $p \to q$ is false,
    Then if $r \to s$ is false,
        (a) $q \lor s$ is false.
        (b) Stop.
5   If $r \to s$ is true,
    Then if $p \lor r$ is true,
        (a) $q \lor s$ is true.
        (b) Stop.
6. $q \lor s$ is undecided.
7. Stop.

The flow chart, and algorithm for Constructive Dilemma (C.D.) all illustrate the relationship between $p \to q$, $r \to s$, $p \lor r$ and $q \lor s$. If the first three are true, the $q \lor s$ statement must be true. If $p \to q$ and $r \to s$ are both false, then $q \lor s$ must be false. Why is this so? The answer can be seen with a little reflection. $p \to q$ can be false only if p is true and q is false. $r \to s$ can be false only if r is true and s is false. If both q and s are false, then $q \lor s$ must be false.

PROBLEM SET: CONSTRUCTIVE DILEMMA
Demonstrate the validity of the arguments given below.

1.
    1. $C \to D$ $\therefore$ D $\lor$ F
    2. $E \to F$
    3. C $\lor$ E

2.

    1.   A V B   ∴ (A & C) V (B & D)
    2.   A → C
    3.   B → D

3.

    1.   A V B   ∴ (D → F) V (D → F)
    2.   A → C
    3.   C → (D → F)
    4.   B → C

4.

    1.   A V B   ∴ [C & (D → F)] V [C & (D → F)]
    2.   A → C
    3.   C → (D → F)
    4.   B → C

5.

    1.   (F V G) & (H → I)   ∴ K V O
    2.   (F → K) & (L → M)
    3.   (G → 0) & (P → Q)

6.

    1.   F → (G V H)   ∴ (L & P) V J
    2.   (I → K) → (H → J)
    3.   (M V O) → [G → (L & P)]
    4.   I → K
    5.   F
    6.   M V O

7.

    1.   R → (S V T)   ∴ F V Z
    2.   U → (W V X)
    3.   (~R & ~U) → (T V Y)
    4.   T → F
    5.   ~(S V T)
    6.   Y → Z
    7.   ~(W V X)

8.

   1. $A \rightarrow [(B \rightarrow C) \& (E \rightarrow F)]$ ∴ C V F
   2. $D \rightarrow (G \rightarrow H)$
   3. D V A
   4. B V E
   5. $(G \rightarrow H) \rightarrow I$
   6. ~I

9.

   1. $J \rightarrow \{[(K \rightarrow L) \& (N \rightarrow O)] \& (P \lor Q)\}$ ∴ L V 0
   2. M V J
   3. $M \rightarrow (R \rightarrow S)$
   4. K V N
   5. $(R \rightarrow S) \rightarrow T$
   6. ~T

10.

   1. $A \rightarrow \{[B \rightarrow (C \rightarrow D)] \& [F \rightarrow (G \rightarrow H)]\}$ ∴ J V K
   2. $E \rightarrow I$
   3. E V A
   4. ~I
   5. B V F
   6. $(C \rightarrow D) \rightarrow J$
   7. $(G \rightarrow H) \rightarrow K$

1. If Augustine is right, then God is eternal in the sense that he transcends all temporal categories. If process philosophers are right, then God is eternal in the sense that he is part of an unending temporal sequence. Either Augustine or process philosophers are right. Therefore, God is eternal in the sense that he transcends all temporal categories or God is eternal in the sense that he is part of an unending temporal sequence.

   A, E1, P, E2

2. If Adam Smith is right, then there is a natural identity of interests in a capitalist economy. If there is a natural identity of interests in a capitalist economy, then capitalism is viable. If Marx is right, then there is a natural conflict of interest in a capitalist economy. If there is a natural conflict of interest in a capitalist economy then a communist revolution is inevitable. Either Adam Smith or Marx is right. Therefore, either capitalism is viable or a communist revolution is inevitable.

    A, I, V, M, C, R.

3. If Social Darwinism is right, then capitalist society favors intelligent and hard working people. But capitalist society exploits intelligent and hard working people if Marxism is right. Either social Darwinism or Marxism is right. Therefore, either capitalist society favors intelligent and hard working people or exploits intelligent and hard working people.

    S, F, M, E.

4. Enrollment doubling in approximately seven years is a necessary condition of enrollment increasing by ten percent each year. Enrollment increasing fifteen percent each year is a sufficient condition for enrollment doubling in five years. Enrollment will increase by either ten percent each year or by fifteen percent each year. Therefore, enrollment is going to double in seven years or in five years.   T, S, F, Y.

5. Enrollment can double in seven years only if we add three English professors in the fifth year. Adding three English professors in the third year is a necessary condition for doubling enrollment in five years. We will double enrollment in seven years or five years. Therefore, we must add three English professors in the fifth year or the third year.

    A, B, C, D.

6. Jones being drunk entails that he is liable. Jones being liable is a sufficient condition for him to pay a fine or go to jail. Jones was home only if he was sober. He was either drunk or home. Therefore, either Jones is liable and he will pay a fine or go to jail, or Jones was sober.

    D, L, F, J, H, S.

7. If Social Darwinism is right, then welfare programs exploit those who are characterized by the virtues of the Protestant work ethic. If welfare programs exploit those who are characterized by the virtues of the Protestant work ethic, then welfare programs interfere with the evolution of virtue. If Social Darwinism is not right, then welfare programs express Christian charity. If welfare programs express Christian charity, then welfare programs promote social justice. Social Darwinism is right or it is not right. Therefore, either welfare programs exploit those who are characterized by the virtues of the Protestant work ethic and interfere with the evolution of virtue, or welfare programs express Christian charity and promote social justice. S, E, I, C, J.

8. Heidi will go to France, or she will go to England or Spain. If she goes to England she will study a semester at Oxford. If she goes to Spain she will study a semester at Madrid. She is not going to France. Therefore, she will study a semester at Oxford or Madrid.   F, E, S, O, M.

9. Lisa will study in Spain one semester or two. If she majors in Spanish, then if she studies in Spain one semester she will complete her Spanish major in the United States. If she also majors in psychology, then if she studies two semesters in Spain she will complete her Spanish major in Valencia. Lisa will major in Spanish. She will also major in psychology. Therefore, Lisa will either complete her Spanish major in the United States or in Valencia. O, T, S, U, P, V.

10. If the appliance was in working order when the house was sold, then the seller is liable for repairs. If the contract specifically guarantees that all appliances be in working order, then the seller is liable for repairs. Either the appliance was in working order when the house was sold or the contract specifically guarantees that all appliances be in working order. Therefore, either the seller is liable for repairs or the seller is liable for repairs. W, L, G.

CHAPTER 16: ADDITION

The principle of addition, abbreviated Add., is symbolized as follows:

Figure 16.1

p		
∴ p	V	q

Figure 16.2

I am a man.
Therefore, I am a man or a snake. M, S.

1.	M		∴ M V S
2.	M V S	1, Add.	

This principle seems to be a curious one until we reflect on the meaning of the "inclusive or" statement. The "inclusive or" affirms only that at least one of the statements linked by the "or" is true. Since it is true that I am a man, the statement which asserts that I am man or that I am anything else in the universe is true. Thus, since it is true that I am a man, it is true that I am a man or a snake. I am a man or a woman. I am a man or I am the planet Mars. That "I am the planet

Mars" is not only false, it is absurd. Nevertheless, the statement that "I am **either** a man or the planet Mars" taken as a whole is a true statement. This principle is depicted below in the form of truth table, a flow chart and an algorithm.

Figure 16.3

TRUTH TABLE FOR ADDITION

p	q	p	→	(p	V	q)
1	1	1	1		1	
1	0	1	1		1	
0	1	0	1		1	
0	0	0	1		0	

Figure 16.4

Flow Chart For Addition

Given the value of p is known, what can be known about the value of p V q?

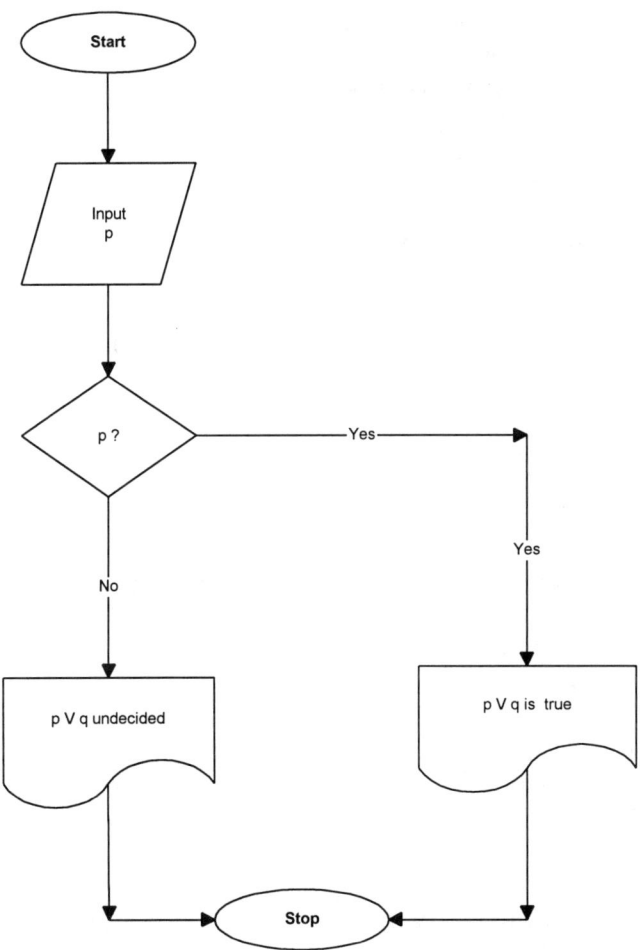

Figure 16.5

Algorithm For Addition

Given the value of p is known, what can be known about the value of
p V q?

    1.   Input the value of p.
    2.   If p is false,
          (a) p V q is undecided.
          (b) Stop.
    3.   p V q is true.
    4.   Stop.

PROBLEM SET: ADDITION

1.
    1.   C   ∴ C V D

2.
    1.   C V E   ∴ F V G
    2.   C → D
    3.   E → F
    4.   ~D

3.
    1.   C V E   ∴ (F V G) V {(P & B) → [D → (R V S)]}
    2.   C → D
    3.   E → (F V G)
    4.   ~D

4.
    1.   C V E   ∴ (F V G) V [D → (C & D)]
    2.   C → D
    3.   E → (F V G)
    4.   ~ D

5.
    1.   {[(A & B) → (C V D)] & (E V F)} → G   ∴ G V H
    2.   E V F
    3.   (A & B) → (C V D)

6.
   1. {[(A & B) & (C V D)] & (E V F)} → G   ∴ G V H
   2. C
   3. A
   4. B
   5. E

7.
   1. (A & B) → J   ∴ K V J
   2. (~C V D) → K
   3. ~C
                 Solve using CD; Solve without using CD

8.
   1. {[(A V B) V (C V D)] V [(E V F) → (G V H)]} → I   ∴ I
   2. A

9.
   1. (A → B) & (C → D)   ∴ F V B
   2. (E → F) & (G → H)
   3. E

10.
   1. [(A → B) & (C → D)] → F   ∴ ~[(A → B) & (C → D)] V ~G
   2. ~F

1. Jones is poor. Therefore, Jones is poor or he is a thief.   P, T.

2. If Jones is delinquent on his taxes or can't meet his payment, then he is in
    serious financial difficulty. Jones is delinquent on his taxes. Therefore,
    Jones is in serious financial difficulty. T, P, F.

3. If Sarah dates Philip or John, then Michael will be angry. Sarah will date
    Philip. Therefore, Michael will be angry or Nancy will be glad.
                             P, J, M, N.

4. If Jones suffers from symptom #1, then he has disease #1 and disease #2.
    Jones does not have both disease #1 and disease #2. Therefore, Jones
    does not suffer from symptom #1 or he does suffer from symptom #2.
                             S1, D1, D2, S2.

5. If John is a minor, then he will not be charged with theft. If John is an adult, then he will certainly go to jail. John is either a minor or an adult. Therefore, John will not be charged with theft or he will certainly go to jail, or he will be charged with murder and will be deported to Florida.
    M, T, A, J, C, D.

6. If Jones is an hourly employee then he gets time and one half for overtime. If he is salaried, then he is on salary schedule S_1. Jones is either an hourly employee or he is salaried. Jones does not get time and one half for overtime. Therefore, Jones is either on salary schedule S1 or S2.
    H, T, S, S1, S2.

7. If Cassie understands Bill in some sense of understand, then computers can think in some sense of think. If Cassie does not understand Bill in some second sense of understand, then computers may not be able to think in some second sense of think. Cassie either understands Bill in some sense of understand or does not understand Bill in some second sense of understand. Therefore, computers either can think in some sense of think or computers may not be able to think in some second sense of think, or computers may be able to think in some third sense of think or computers may not be able to think in some third sense of think.
    U1, T1, U2, T2, T3.

8. If Spinoza is right, then emotions are confused ideas. Emotions are not confused ideas. Therefore, Spinoza is not right or love is possible for God only if God is not a perfectly intelligent being.   S, C, L, I.

9. Emotions are identical with confused ideas only if computer programs having confused ideas is a sufficient condition for computer programs to have emotions. Emotions are identical with confused ideas and Cassie is a computer program. Computer programs can have confused ideas. Therefore, computer programs can have emotions or I am confused about computer programs.   I, H, E, C, A.

10. If the following are true, then Cassie can think: Cassie can compute, or Cassie has emotions and experiences guilt.   Cassie can compute. Therefore, Cassie can think.   T, C, E, G

CHAPTER 17: MATERIAL IMPLICATION, DEDUCTIVE ARGUMENTS, AND ABBREVIATED TRUTH TABLES

When we develop a valid deductive argument from the rules of inference and stated premises, what are we asserting? We assert that, given a set of premises and certain rules of inference that a given conclusion must necessarily follow. Or to put it another way, we **deny** the possibility that all of the premises stated could be true and the conclusion false. We affirm that if the premises are true, then the conclusion necessarily follows. In other words, we assert that there is a relationship of material implication between the premises and the conclusion of the argument.

We can illustrate this by consideration of the following valid deductive argument. First, we will use the rules of inference with which we have been working. Secondly, we will show the argument to be valid by use of truth tables.

Figure 17.1

If Jones is guilty, then he will be convicted. Jones is either innocent or guilty. Jones is not innocent. Therefore, Jones will be convicted. G, C, I.

1.	$G \rightarrow C$	$\therefore$ C
2.	I V G	
3.	~I	
4.	G	2, 3 D.S.
5.	C	1, 4 M.P.

Truth Table For Valid Deductive Argument Given Above

Figure 17.2

	G	C	I	{[(G	→	C)	&	(I	V	G)]	&	~I}	→	C
													p	q
1	1	1	1			1					0	0	1	1
2	1	1	0			1					1	1	1	1
3	1	0	1			0					0	0	1	0
4	1	0	0			0					0	1	1	0
5	0	1	1			1					0	0	1	1
6	0	1	0			0					0	1	1	1
7	0	0	1			1					0	0	1	0
8	0	0	0			0					0	1	1	0

In the truth table above we have taken the conjunction of all the premises given in the argument and asserted that all of these premises taken together imply the conclusion C (for Jones will be convicted). When we assign all of the truth values to the variables, we see that there is no instance where the conjunction of the premises is true and the conclusion is false. This results in the final column of our truth table containing all 1's.

It is clear from this that if we chose, we could dispense with rules of inference in favor of truth tables. It is equally clear that this would be very cumbersome. If we were dealing with a problem using six variables this would require a truth table with sixty four columns. Seven variables would require one hundred and twenty-eight columns.

Abbreviated Truth Tables

There is a short cut to our use of truth tables. The premises of our argument can fail to entail the conclusion **C** only under those conditions where **C** is false. This is because **C** is functioning as our **q** statement and the premises as our **p** statement. **p → q** is false when and **only when p** is true and **q is false**. Thus, if C (our q statement) is true, then it does not matter whether the conjunction of the premises (our **p** statement) is true or false. **p → q** will always be true when **q** is true. When we review our complete truth tables in Figure 17.2, we see that **C** is false in only four instances, logical possibilities (3), (4), (7) and (8). We thus can safely ignore all of the other possibilities as irrelevant to demonstrating that our argument is invalid or valid.

One way of proving the validity of an argument is to attempt by a **systematic and complete** method to prove the argument to be invalid. If this systematic and complete attempt to prove the argument invalid fails, then we know the argument to be valid. Let us illustrate this method by attempting to prove our present argument invalid. We know that the only way our argument is invalid is if C is false. Let us begin by making C false.

Figure 17.3

{[(G	→	C)	&	(I	V	G)]	&	~I}	→	C
										0

But if C is false in the conclusion, then it must be false everywhere it appears in the premises.

Figure 17.4

{[(G	→	C)	&	(I	V	G)]	&	~I}	→	C
		0								**0**

Now having made C (our q statement) false, let us try to make the argument invalid by proving the premises (our p statement) true. Our best strategy seems to be to work from left to right in attempting to prove all the component parts of the premises true. In order for every-thing between the outer brackets { } to be true that portion between the next inner brackets [ ] must be true. This requires that **both** (G → C) and (I V G) must be true. If G is true, then (G → C) would be false G → C).

   1 0 0

Thus, in order to **save** G → C, we will make G false as in Figure 17.5.

Figure 17.5

{[(G	→	C)	&	(I	V	G)]	&	~I}	→	C
0		0								
	1									**0**

But we must make G false through out. This we will do in Figure 17.6.

Figure 17.6

{[(G	→	C)	&	(I	V	G)]	&	~I}	→	C
0	1	0				0				0

In order to **save** the entire phrase [ ] (I V G) must be true. But for I V G to be true, I must be true. We will thus make I true in Figure 17.7.

Figure17.7

{[(G	→	C)	&	(I	V	G)]	&	~I}	→	C
0	1	0		1	1	0				0
			1							

Thus, we have **saved** the first phrase but in so doing we had to make I true. But this means that ~I must be false. We now make I False in Figure 17.8.

Figure 17.8

{[(G	→	C)	&	(I	V	G)]	&	~I}	→	C
0	1	0	1	1	1	0		0		0

But if ~I is false then the phrase taken as a whole { } is false. This follows from that fact that the conjunction of a true and a false statement is false. This results in a statement which asserts that the premises entails the conclusion being true. This is shown in Figure 17.9 where we now have a false p statement entailing a false q statement.

Figure 17.9

{[(G	→	C)	&	(I	V	G)]	&	~I}	→	C
0	1	0	1	1	1	0		0 1		0
							0		**1**	

We have in a systematic and complete way (in that we exhausted all of the possibilities) tried to prove the premises true simultaneously with the conclusion being false. **Our failure, in the light of our systematic and thorough attempt, proves that the argument is valid.**

Not all uses of the abbreviated method of truth tables are this simple. For example, let us suppose that the conclusion of the argument were (G & C). There are **three** possible ways in which G & C could be false corresponding to logical possibilities number 2, number 3 and number 4. These are illustrated below in Figure 17.10..

Figure 17.10

	G	&	C
1.	1	1	1
2.	1	0	0
3.	0	0	1
4.	0	0	0

To prove that the premises could not be true while the conclusion is false, we have to show that **none** of the three possible ways in which G & C could be false are compatible with the premises being true. We will demonstrate the validity of our argument by taking these possibilities one at a time.   Figure 17.11 takes logical possibility number 2 and treats G as false and C as true.

Figure 17.11

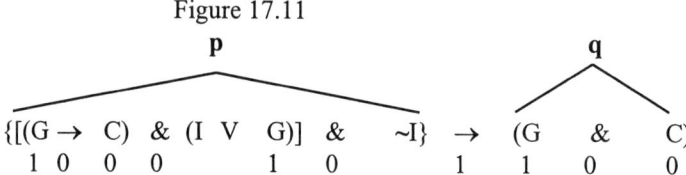

{[(G → C) & (I V G)] & ~I} → (G & C)
 1 0   0   0   1   0   1   1 0   0

This first way in which G & C may be false (which we have designated as logical possibility number 2) passes the test of validity. Since G is true and C is false, then G → C is false, and the conjunct of (G → C) & (I V G) must be false. If this is the case, however, the conjunct of [(G → C) & (I V G)] and ~I must be false. Thus, our p statement is false and therefore entails our q statement.

This, however, is not sufficient. We have to check out the other possibilities which we will proceed to do. We thus must proceed to logical possibility number 3 in which G is false and C is true. This is shown in Figure 1712.

Figure 17.12

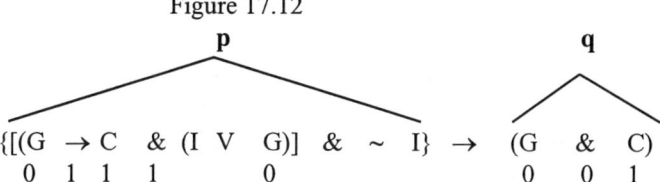

$$\{[(G \rightarrow C \quad \& \ (I \ V \quad G)] \quad \& \quad \sim \ I\} \rightarrow \quad (G \quad \& \quad C)$$
$$\phantom{\{[(G} 0 \quad 1 \ 1 \quad 1 0 \rightarrow (} 0 \quad 0 \quad 1$$

The explanation for our second possibility is only slightly more complicated. Since, G is false and C is true we first fill in truth values for these variables in our premises. We also observe that this makes G → C true as is shown in Figure 17.13.

Figure 17.13

$$\{[(G \rightarrow C) \quad \& \ (I \ V \ G)] \quad \& \quad \sim \ I\} \rightarrow \ (G \ \& \quad C)$$
$$\phantom{\{[(}0 \quad 1 \ 1 0 \rightarrow (}0 \ 0 \quad 1$$

In the process, however, we have made G false which means that in order to save the I V G statement we must make I true. This enables us to save our first conjunct as in Figure 17.14.

Figure 17. 14

$$\{[(G \rightarrow \ C) \quad \& \quad (I \quad V \ G)] \ \& \quad \sim \ I\} \rightarrow \ (G \ \& \ C)$$
$$\phantom{\{[(}0 \ 1 \quad 1 \quad 1 \ 1 \ 1 \ 0 \quad 0 \quad 0 \ 1 \quad 1 \quad 0 \ 0 \ 1$$

We have saved the first conjunct but in making I true we have been forced to make ~I false. This makes the conjunction of ~I and the rest of the p statement false. Thus, once again our p statement is false and therefore entails our q statement. We cannot, however, affirm the validity of our argument until we have checked out the final possibility which we have designated as logical possibility number 4 from Figure 17.10.

Figure 17.15

$$\{[(G \rightarrow \ C) \quad \& \quad (I \quad V \ G)] \ \& \quad \sim \ I\} \rightarrow \ (G \ \& \ C)$$
$$\phantom{\{[(}0 \ 1 \quad 0 \quad 1 \quad 1 \ 1 \ 0 \quad 0 \quad 0 \ 1 \quad 1 \quad 0 \ 0 \ 0$$

Figure 17.15 summarizes this effort with results similar to the other two attempts G and C are both false. This makes G → C true. So far so good. But if G is false then in order to make (I V G) true we must make I true. But if I is true then ~I is made false. The conjunction of ~I with [(G → C) & (I V G)] is thus false. Since our entire p statement which is {[(G → C) & (I V G)] & ~I} is false the statement {[(G → C) & (I V G) & ~I)]} → (G & C) is true.

Thus, we have checked out <u>all three</u> possible ways for our q statement to be false and found that in none of those situations could we make our p statement to be true. We have thereby proved the validity of the argument by the method of an abbreviated truth table.

This is obviously still an improvement over the non-abbreviated method. Our complete table had eight columns. Our abbreviated truth table, on the other hand, had only three columns. This is shown below in figure 17.16.

Figure 17.16

{[(G	→	C)	&	(I	V	G)]	&	~	I}	→	(G	&	C)
1	0	0	0	1	0	1	0	0	1	1	1	0	0
0	1	1	1	1	1	0	0	0	1	1	0	0	1
0	1	0	0	0	0	0	0	1	0	1	0	0	0

But suppose the conclusion were [(G & C) & (R → S)]. There are thirteen different ways (out of sixteen possibilities) as shown in figure 17.17 in which this conclusion could be false (possibilities 2, 8, and 5-13). Thus, even the abbreviated method of truth tables can in certain circumstances be very cumbersome.

Figure 17.17

Truth Table For (G & C) & (R → S)

	G	C	R	S	(G	&	C)	&	(R	→	S)
1	1	1	1	1	1	1	1	**1**	1	1	1
2	1	1	1	0	1	1	1	**0**	1	0	0
3	1	1	0	1	1	1	1	**1**	0	1	1
4	1	1	0	0	1	1	1	**1**	0	1	0
5	1	0	1	1	1	0	0	**0**	1	1	1
6	1	0	1	0	1	0	0	**0**	1	0	0
7	1	0	0	1	1	0	0	**0**	0	1	1
8	1	0	0	0	1	0	0	**0**	0	1	0
9	0	1	1	1	0	0	1	**0**	1	1	1
10	0	1	1	0	0	0	1	**0**	1	0	0
11	0	1	0	1	0	0	1	**0**	0	1	1
12	0	1	0	0	0	0	1	**0**	0	1	0
13	0	0	1	1	0	0	0	**0**	1	1	1
14	0	0	1	0	0	0	0	**0**	1	0	0
15	0	0	0	1	0	0	0	**0**	0	1	1
16	0	0	0	0	0	0	0	**0**	0	1	0

CHAPTER18: MATERIAL IMPLICATION, DEDUCTIVE ARGUMENTS, ABBREVIATED TRUTH TABLES AND THE TEST FOR CONSISTENT PREMISES

There is, however, at least one other consideration which must be taken into account here. $p \rightarrow q$ can be false **only if** the premises are true and the conclusion is false. But what if we had an argument with contradictory premises such as follows:

Figure 18.1

1.	$A \rightarrow B$	$\therefore B$
2.	A	
3.	~B	

This would seem to allow for proof by our rules of inference. For example:

Logic of the Computing Sciences

Figure 18.2

1.	A → B		∴ B
2.	A		
3.	~B		
4.	B	1, 2 M.P.	

But something is clearly wrong here. Not only can we prove B but since we also have ~B in premise three, we can prove B and ~B as in step five:

5.	B & ~B	4, 3 Conj.

Any argument which allows one to prove a contradiction is surely an embarrassment. But how would we fair if we used our truth table method?

Figure 18.3

{[(A	→	B)	&	A]	&	~	B}	→	B
0	1	0	0	0		1	0		0
					0			**1**	

Our abbreviated truth table method shows the argument to be valid. **Having made the conclusion false**, we are unable to make the premises true. Thus, we must judge the argument as valid. Does this demonstrate that the truth table method for demonstrating validity is inadequate.   Yes and no!   It is adequate but only with some modification. **Having made the conclusion false**, we are unable to make the premises true. But as a matter of fact, **it is impossible to make the premises true under any circumstances**. This is because the premises are contradictory or we may say inconsistent. This is demonstrated below:

Figure 18.4

{[(A	→	B)	&	A]	&	~	B}
1	1	1	1	1	0	0	1

In order to save the first phrase [   ] we must make A true so that the conjunction of A and (A → B) can be true. But this requires us to make B true so that A → B may be true. But if B is true, then ~B must be false and if ~B is false then, of course, the conjunction of [(A → B) & A] & ~B cannot be true.

Thus, we judge {[(A → B) & A] & ~B} → B to be an invalid argument **by reason of inconsistent premises**.

Therefore, a complete test of the validity of an argument by truth tables requires the following:

1.   Test the premises of the argument to see if they are capable of being true under <u>any</u> circumstances.
2.   **If the answer is no, then designate the argument as invalid by reason of inconsistent premises.**
3.   If the answer is yes, then test to see if the premises can be made true, when the conclusion is made false (under any of the circumstances under which the conclusion can be made false).
4.   If the answer is yes, then designate the argument as invalid.
5.   If the answer is no, then designate the argument as valid.

This can be illustrated by both a flow chart and an algorithm analogous to those used in computer science programming.

Figure 18.5

Flow Chart For Abbreviated Truth Tables Method For Proving Validity Or Invalidity. (Assumes three possible ways of making conclusion false)

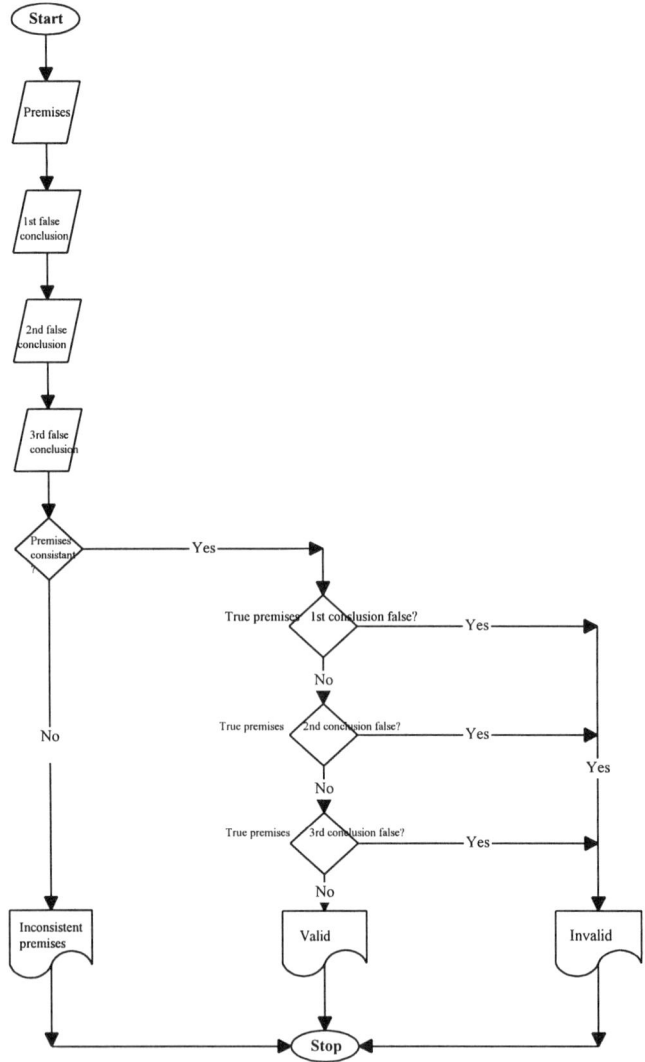

Figure 18.6

Algorithm For Abbreviated Truth Tables Method For Proving Validity Or Invalidity (assuming three possible ways of making conclusion false)

1. Input premises
2. Input 1st way of making conclusion false.
3. Input 2nd way of making conclusion false.
4. Input 3rd way of making conclusion false.
5. If premises are inconsistent,
     (a) Invalid by reason of inconsistent premises.
     (b) Stop.
6. If premises can be made true when conclusion is false in the first way that conclusion can be made false,
     (a) Argument is invalid.
     (b) Stop.
7. If premises can be made true when conclusion is false in the second way that conclusion can be made false,
     (a) Argument is invalid.
     (b) Stop.
8. If premises can be made true when conclusion is false in the third way that conclusion can be made false,
     (a) Argument is invalid.
     (b) Stop.
9. Argument is valid.
10. Stop.

We will now write some algorithms for particular arguments which include a test for inconsistent premises. These should be taken merely as illustrations. In any concrete situation the specific detail which will be required will depend on the nature of the problem and on the logic which is built into the specific language and system which you are using. Furthermore, it is possible to write several different algorithms for the same problem which are equally correct.

We will start with an algorithm for the argument which we have previously observed to be inconsistent.

Figure 18.7

1.	A → B	∴ B
2.	A	
3.	~ B	

This may be written as follows: {[(A → B) & A] & ~ B} → B

Algorithm for {[(A → B) & A] & ~ B} → B

1.   Input value of A → B [1] as true.
2.   Input value of A as true.
3.   Input value of B as false.
4.   If A → B is consistent with A,
      Then if [~(A & ~B) & A] is consistent with ~B,
      go to line #6.
5.   (a) Invalid by reason of inconsistent premises.
      (b) Stop.
6.   If B is false,
      Then if {[~(A & ~B) & A] & ~B is true,
            (a) Argument is invalid.
            (b) Stop.
7.   Argument is valid.
8.   Stop

Explanation

Our first step is to put in all our premises as true. Since A → B excludes the logical possibility of A being true while B is false it can be written ~(A & ~B). In line #4 we test for the consistency of the premises. If the premises are inconsistent we go to line #5 and the output "invalid by reason of inconsistent premises." If the premises are consistent, then the algorithm is instructed to go to line #6 and a check for validity. If the problem passes the validity check we get the output of line #7.

---

[1] Note that A → B is logically equivalent to ~(A & ~B) which can therefore be substituted for A → B. This makes the algorithm easier to construct.

Our second algorithm will be a little more complicated than the previous one since the argument allows the conclusion to be false under three different conditions.

Figure 18.8

1. $A \rightarrow B$ $\therefore$ B & C
2.   A
3.   ~C

Algorithm for $\{[(A \rightarrow B) \& A] \& \sim C\} \rightarrow (B \& C)$

1.   Input value of A $\rightarrow$ B) as true.
2.   Input value of A as true.
3.   Input value of C as false.
4.   If A $\rightarrow$ B is consistent with A,
        Then (a) if [A $\rightarrow$ B) & A] is consistent with ~C,
        go to line #6.
5.   (a) Invalid by reason of inconsistent premises.
        (b) Stop.
6.   If (~B & C) is true,
        then if $\{[\sim(A \& \sim B) \& A] \& \sim C\}$ is true,
                (a) Argument is invalid.
                (b) Stop.
7.   If (B & ~C) is true,
        then if $\{[\sim(A \& \sim B) \& A] \& \sim C$ is true,
                (a) Argument is invalid.
                (b) Stop.
8.   If (~B & ~C) is true,
        then if $\{[\sim(A \& \sim B) \& A] \& \sim C\}$ is true,
                (a) Argument is invalid.
                (b) Stop.
9.   Argument is valid.
10.   Stop.

Explanation

The explanation here is similar to the explanation for our previous algorithm. The significant difference is that there are three ways that the B & C can be false. We must check each of these three ways before we can prove an argument valid.

If there were ten or twenty ways the conclusion could be false, we might be lucky enough to prove the argument invalid on our first attempt. We might prove it invalid on the tenth or twentieth attempt. We could only prove it valid, however, by testing **all** the possibilities.

PROBLEM SET:

1. A → B can be written ~(A & ~B). It may also be written {[(A & B) V (~A & B)] V (~A & ~B)}. Explain why both of these interpretations are correct. What would be the implication for writing an algorithm if the second interpretation were used?

2. Write an algorithm for the following problem.

    (1) A → B   ∴ B & ~C
    (2) A → C
    (3) A

3. Write an algorithm for the following problem.

    (1) A V C   ∴ C V D
    (2) ~A
    (3) B

CHAPTER 19: MATERIAL IMPLICATION
AND
SCIENTIFIC THEORIES

The notion of material implication gives us an interesting and instructive way of looking at scientific theories. Typically a scientific theory asserts a number of propositions concerning relationships in the empirical world that are said to exist under certain specified conditions. Given those conditions and relationships we deduce certain events from the theory. We test the theory by empirical tests which are designed to determine if the events or states of affairs deduced from the theory actually exist in the empirical world. If they do not exist then we take this as evidence that the theory is at least partially false. If the states of affairs deduced from the theory do exist in the world then we take this as evidence confirming the theory. This last assumption, however, is not strictly true. Philosophers of science have observed that scientific theories can be disproved but **they cannot conclusively be proved.** The reason for this assertion can be readily understood if we see scientific theories as standing in a relationship of material implication to their conclusions.   If given certain initial conditions a theory (T) implies that certain events (E) will occur, then we may illustrate it as below:

Figure 19.1

	T	→	E
(1)	1	1	1
(2)	1	0	0
(3)	0	1	1
(4)	0	1	0

As we can see, if the conclusion is shown empirically to be false then the theory (at least in part) is shown to be false. (Logical possibility number (2). But a true conclusion is compatible with a false theory (logical possibility number (3). Thus, the fact that we have empirical verification of the conclusions of a theory does not necessarily establish the truth of the theory. Of course, it can be argued that repeated testing of a theory tends to provide evidence that the theory is probably true, since if it were not, the probability of logical possibility number (2) occurring in repeated experiments is high.

Looked at in this way, the refutation of a scientific theory may be viewed as a variation on Modus Tollens. As is shown in figure 19.2.

Figure 19.2

	T	→	E
	~	E	
∴	~	T	

This of course is an over simplification and requires some qualification and expansion. In the first place, many scientific theories are cast in the form of inductive rather than deductive arguments. In other words, what is claimed is not that if the theory (T) is true that the conditions (E) will certainly follow but merely that E will <u>probably</u> follow or follows a certain percentage of the time. An inductive argument may be modified to take on a deductive form by making a claim such as the following. Given certain initial conditions C1, 2, 3, ... 8, T entails that E will occur X percentage of the time. Given this specification, one failure of the occurrence of E would not refute T, but repeated failures of the occurrence of E would constitute a refutation.

Secondly, scientific theories are usually complex entities composed of many parts. Thus, T may equal (T1 & T2 & T3 & . . . . . . T10). It also is likely to be defined to work under a considerable number of specific conditions. Thus, when events predicted by a theory fail to occur, rather than throwing out the theory entirely, scientists will most frequently modify the theory either by adjusting the specifications for the initial conditions or some aspects of the theory. There is obviously a great deal more to be said about the nature of scientific theories and their confirmation or refutation.

Further discussion, however, would take us far beyond the appropriate task of this text. What needs to be pointed out here is that the basic form of scientific theories is in the form of material implications which define specific conditions, specific theoretical relationships and assert that they entail specific events as is illustrated in figure 19.3.

Figure 19.3

$$\{[C1 \text{ \& } C2 \text{ \& } C3 \text{ \& } C4, \text{ \& } C5] \text{ \& } [T1 \text{ \& } T2 \text{ \& } T3 \text{ \& } T4 \text{ \& } T5]\} \rightarrow (E1 \text{ \& } E2 \text{ \& } E3 \text{ \& } E4 \text{ \& } E5)$$

CHAPTER 20: THE NATURE OF MATERIAL EQUIVALENCE

We are now in a position to understand several other important concepts related to our study. **Two variables are said to be materially equivalent when they are simultaneously both true or both false.** The symbol of equivalence is the double arrow (↔). Thus, if we assert that p is equivalent to q we would write and develop the truth table as below.

Figure 20.1

	p	q	p	↔	q
(1)	1	1	1	1	1
(2)	1	0	1	0	0
(3)	0	1	0	0	1
(4)	0	0	0	1	0

The statement that p is equivalent to q is true in logical possibilities number (1) (where **both** p and q are true) and logical possibilities number (4) (where **both** p and q are false). In logical possibilities number (2) and number (3) it is false.

Another way of saying this is that p is a **bi-conditional** of q. In ordinary English we can say p, if and only if, q. This points up

another feature of the bi-conditional. p ↔q is equivalent to, if p then
q **and** if q then p.   This can also be shown by means of a truth table.

Figure 20.2

P	q	(p	↔	q)	↔	[(p	→	q)	&	(q	→	p)]
1	1	**1**		1	1	1	1	1	**1**	1	1	1
1	0	**0**		1	1	0	0	**0**	0	1	1	
0	1	**0**		1	0	1	1	**0**	1	0	0	
0	0	**1**		1	0	1	0	**1**	0	1	0	

As can be seen from the above truth table, p is a bi-conditional for
q, and p entails q and q entails p, have identical truth tables which
demonstrates that they are **logically equivalent.**

This introduces the notion of logical equivalence and of tautology.
Two statements are **logically equivalent** when the assertion of their
equivalence is a **tautology**. What is a tautology? A tautology is a
statement which is true by definition. A red hat is red is a tautological
statement. A little reflection will reveal that this statement does not
directly tell us anything about any red hats in the world. It does not
even tell us whether or not there are any red hats in the world. It
simply tells us how we have defined terms in the English language. In
this it resembles "All circles are round." or "A unicorn is a one horned
animal." The last statement we accept to be true, even though we
believe that there is no such thing as a unicorn in the empirical world.
That "All circles are round" is true by virtue of the way we define
"circle" and "round." This is obvious to any competent user of the
English language. The valid conclusions of a logical system are also
true by definition of the meaning of the terms and the rules of
inference we have designated. That the conclusions are necessarily
true, however, is not always immediately obvious.   Likewise, the
output of a computer is **true by definition** given the structure of the
program and the inputs.   These outputs are, of course, not always
obvious. If they were **obviously** true, computers would not be as useful
as they are.

That p ↔ ⁓p is obvious on a limited amount of reflection. It is
necessarily the case (true by definition that if p is true then ⁓ ⁓ p will

be true and that if p is false that ~ ~ p will be false. This is because p and ~ ~p are **logically equivalent.** This can be shown by a truth table.

Figure 20.3

p	↔	(~	~	p)
1	1	1	0	1
0	1	0	1	0

It may not be as immediately obvious but p → q is logically equivalent to ~p V q.

Figure 20.4

	p	q	(p	→	q)	↔	~	p	V	q
(1)	1	1	1	1	1	1	0	1	1	1
(2)	1	0	1	0	0	1	0	1	0	0
(3)	0	1	0	1	1	1	1	0	1	1
(4)	0	0	0	1	0	1	1	0	1	0

Examination of the truth table shows that p → q and ~ p V q are true for logical possibilities number (1), (3) and (4) and false for logical possibility number (2). They are therefore always true and false (by definition) under the same conditions. This makes them logically equivalent.

Rules of Replacement

If two statements are logically equivalent, then for purposes of solving a logical problem one statement can **replace** the other with no loss of meaning. This is true of computer programs and of legal documents. It is frequently not the case, however, in poetry and many types of prose where consideration of grace, style, convention and emotive effect may be as important as logical syntax. It is clear from our previous discussion that using the truth table method we would be able to derive a set of Rules of Inference and a set of Rules of Replacement of our own. For purposes of this course, however, we will accept the list which has come to be accepted as standard by modern logicians. There are a number of reasons for this. In the first place, we would probably, if left to our own devices, come up with a list which was in many respects similar to the list developed already by others. Secondly, although we could develop a logical system with twenty five or perhaps a hundred rules of replacement, the standard logic with nine rules of inference and ten rules of replacement is adequate to solve all of the problems we are likely to confront.[*] Since these rules have to be understood, memorized and then worked with, a shorter list is desirable. Below you are given a list of the nineteen rules.

RULES OF INFERENCE

1.	**Modus Ponens** (M.P.)	2.	**Modus Tollens** (M.T.)	3.	**Hypothetical Syllogism** (H.S.)
	$p \rightarrow q$ p		$p \rightarrow q$ $\sim q$		$p \rightarrow q$ $q \rightarrow r$
$\therefore$	q	$\therefore$	$\sim p$	$\therefore$	$p \rightarrow r$

4.	**Disjunctive Syllogism** (D.S.)	5.	**Constructive Dilemma** C.D.)	6	**Absorption** (Abs.)
			$(p \rightarrow q) \,\&\, (r \rightarrow s)$		
	$p \lor q$ $\sim p$		$p \lor r$		$p \rightarrow q$
$\therefore$	q	$\therefore$	$q \lor s$	$\therefore$	$p \rightarrow (p \,\&\, q)$

7.	**Simplification** (Simp.)	8.	**Conjunction** (Conj.)	9.	**Addition** (Add.)
			p		
	$p \,\&\, q$		q		p
$\therefore$	p	$\therefore$	$p \,\&\, q$	$\therefore$	$p \lor q$

---

[*] For the time being. We will introduce some additional rules of inference and replacement later on in this text.

RULES OF REPLACEMENT

Any of the following logically equivalent expressions can replace each other wherever they occur:

10.	Double Negation (D.N.):	$p \leftrightarrow \sim \sim p$
11.	Commutation (Com.):	$(p \lor q) \leftrightarrow (q \lor p)$
		$(p \& q) \leftrightarrow \ll (q \& p)$
12.	Tautology (Taut.):	$p \leftrightarrow (p \lor p)$
		$p \leftrightarrow (p \& p)$
13.	Association (Assoc.):	$[p \lor (q \lor r)] \leftrightarrow [p \lor q) \lor r]$
		$[p \& (q \& r)] \leftrightarrow [p \& q) \& r]$
14.	Transposition (Trans.):	$(p \rightarrow q) \leftrightarrow (\sim q \rightarrow \sim p)$
15.	Material Implication (Impl.):	$(p \rightarrow q) \leftrightarrow (\sim p \lor q)$
16.	Exportation (Exp.):	$[p \& q) \rightarrow r] \leftrightarrow [p \rightarrow (q \rightarrow r)]$
17.	Material Equivalence (Equiv.):	$(p \leftrightarrow q) \leftrightarrow [(p \rightarrow q) \& (q \rightarrow p)]$
		$(p \leftrightarrow q) \leftrightarrow [(p \& q) \lor (\sim p \& \sim q)]$
18.	Distribution (Dist.):	$[p \& (q \lor r)] \ll [(p \& q) \lor (p \& r)]$
		$[p \lor (q \& r)] \ll [(p \lor q) \& (p \lor r)]$
19.	De Morgan's Theorems (De M.):	$\sim(p \& q) \leftrightarrow (\sim p \lor \sim q)$
		$\sim(p \lor q) \leftrightarrow (\sim p \& \sim q)$

The ten rules of replacement are intuitively obvious upon some reflection. Some of them, however, are more transparently so than others. The author has generally speaking introduced them by giving the most obvious first and the less obvious later on. I say generally speaking, however, since what is most "obvious" will vary somewhat from individual to individual. As with the rules of inference you will need to memorize all of the rules of replacement eventually. The best plan, however, is to start not by attempting to memorize the rule but by attempting to understand the principle. Two suggestions will be helpful here. The rules of replacement will be introduced one at a time and you are encouraged to demonstrate for yourself by means of truth tables that the expressions noted are in fact logically equivalent. Secondly, you will be given a list of problems which require you to use in their solutions principles learned previously as well as the principle most recently introduced. **The best way to fix the principles in one's mind is to us them in the solution of problems.**

CHAPTER 21:   DOUBLE NEGATION

$$p \leftrightarrow \mathord{\sim}\mathord{\sim} p$$

The rule of double negation or D.N. is so intuitively obvious that it scarcely needs any explanation. It is not the case that I do not have a car is a more cumbersome, but none the less logically equivalent way of saying that I do have a car. The various ways in which this principle functions in solving problems can best be comprehended by applying them to particular problems. Consider the following examples:

Figure 21.1

1.

1.	$D \rightarrow \mathord{\sim} A$		$\therefore \mathord{\sim} D$
2.	A		
3.	$\mathord{\sim}\mathord{\sim} A$	2, D.N.	
4.	$\mathord{\sim} D$	1, 3 M.T.	

Figure 21.2

2.

1.	$(\mathord{\sim}\mathord{\sim} D \rightarrow E) \rightarrow F$		$\therefore F$
2.	$D \rightarrow E$		
3.	$\mathord{\sim}\mathord{\sim} D \rightarrow E$	2, D.N.	
4.	F	1, 3, M.P.	

In the first example, in order to work a modus tollens on premise one to derive ~ D, we needed the negation of our q statement. Since the q statement is ~ A its negation would be ~(~ A). Since A which is given in premise #2 is logically equivalent to ~ ~ A we can derive ~ ~ A from A by the principle of replacement known as double negation and then work the modus tollens to derive ~ D from 1, 3 M.T. In problem number two, we have a some what different problem. In order to derive F we need to work a modus ponens on premise #1. In order to do this we need the p statement which is ~ ~ D → E. But D → E which is given in premise #2 is not strictly identical with ~ ~ D → E. There is, however, a **logical** identity. By using the principle of D.N. on premise #2 we are able to derive ~ ~ D → E. Since premise #2 and the derived premise #3 are logically identical we are able to make a legitimate substitution and solve our problem.

Figure 21.3

Truth Table For Double Negation

p	↔	~	~	p
1	1	1		
0	1	0		

Figure 21.4

Flow Chart For Double Negation

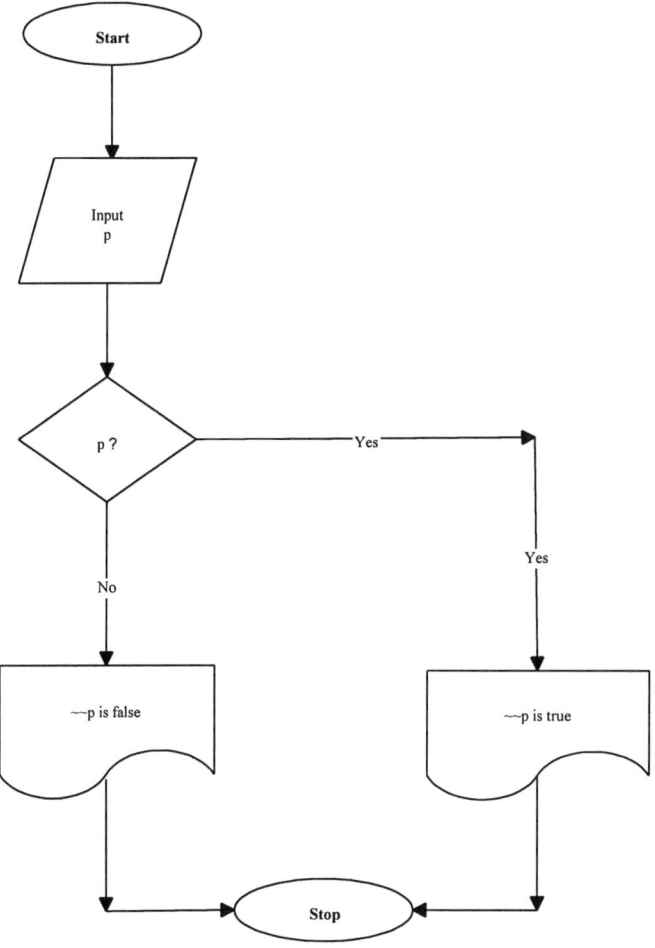

Figure 21.5

Algorithm For Double Negation

1.   Input value of p
2.   If p is true,
      (a) ~ ~ p is true.
      (b) Stop.
3.   ⁓ p is false.
4.   Stop.

PROBLEM SET: DOUBLE NEGATION
Demonstrate the validity of the following arguments.

3.
   1.   (~ ~ D → E) → F   ∴ ~ ~ F
   2.   D → E

4.
   1.   D → ~ ~ A   ∴ ~ ~ ~ D
   2.   ~ A

5.
   1.   D → E   ∴ ~D & ~ ~ ~ F
   2.   ~ ~ ~ E
   3.   F → ~ G
   4.   G

6.
   1.   ~ ~ (D → E) → (~ ~ F & ~ G)   ∴ F V H
   2.   ~ ~ ~ ~ (~ ~ D → E)

7.
   1.   H → (I & J)   ∴ ⁓ H → [H & (I & ⁓ J)]

8.
   1.   ~ ~ A → C   ∴ ~ ~ (D → F) V (D → ~ ~F)
   2.   A V ~ ~ B
   3.   C → (~ ~ D → F)
   4.   B → ~ ~ C

9.

    1. ~H → I   ∴ K
    2. H → (~ J V K)
    3. J
    4. ~ I

10.

    1. ~H   ∴ ~D
    2. F → H
    3. ~A → B
    4. C
    5. ~B
    6. A → [(~C V (D → F)]

1.   If it isn't the case that Rawls is right, then its not the case that the social contract defense of ethics is adequate. It just isn't true that the social contract defense isn't adequate. Therefore, Rawls is right. R, S.

2.   This law is just only if it is rational. I am not willing to make this law universal, only if it is not rational. This law is just. Therefore, I am willing to make this law universal. J, R, U.

3.   If I am rational, then I am autonomous. Being autonomous is a sufficient condition for being moral. If I don't treat others as ends in themselves, then I don't obey the golden rule. Obeying the golden rule is a necessary condition for being moral. Cheating entails not treating others as ends in themselves. Unfortunately, it is not the case that I do not cheat. Therefore, I am not rational and I am not autonomous. R, A, M, E, G, C.

4.   If I am not willing to make the principle of this action into a universal law, then it is not rational. The principle of this action is not rational only if it is not moral. The principle of this action is moral. Therefore, I am willing to make it into a universal law. U, R, M.

5.   If this principle is not one I would be willing to choose were I completely ignorant of my social situation, then it is not one I am willing to make into a universal law. If this principle is not one I am willing to make into a universal law, then it is not a just principle. This is a just principle. Therefore, this principle is one I would be willing to choose were I completely ignorant of my social situation. I, U, J.

6. If it is not the case either that Jones is liable or is able to pay a fine, then this matter cannot be resolved. This matter can be resolved. Therefore, Jones is either liable or is able to pay a fine. L, A, R.

7. If it is not the case either that Hinduism is true or that Buddhism is true, then Islam is true or secular humanism is true. It is false that either Islam or secular humanism true. Therefore, Hinduism or Buddhism is true, or Christianity is true. H, B, I, S, C.

8. Susan working overtime is a necessary condition for Susan to make time and one half pay. Susan will work overtime only if production is not down. Production is down. Therefore, Susan will not get time and one half pay. W, T, P.

9. God is not good only if man is not free. It is not the case both that God is not good and that man is not free. Therefore, God is good. G, F.

10. (A soap opera problem) Suzy's being faithful to Jim is a necessary condition for its not being true that her not being early last night is a sufficient condition for her car having broken down. Its being false that Jane was late only if Jane's car ran out of gas is a sufficient condition for Jane to be faithful to Bill. Suzy is not faithful to Jim. Jane is not faithful to Bill. Either Suzy was not early last night or Jane was late last night. Therefore, either Suzy's car broke down or Jane's car ran out of gas.
    S, E, C, L, G, J.

CHAPTER 22: COMMUTATION

(p V q) ↔ (q V p)
(p & q) ↔ (q & p)

Commutation (abbr. Com.) is another rule of replacement which is intuitively obvious. It is clear that if I have a Ford or a Chevy that it is also true that I have a Chevy or a Ford. Likewise that having both a Chevy and a Ford is logically equivalent to having a Ford and a Chevy.

Commutation thus simply reverses the order or sequence of either p V q to q V p, or p & q to q & p. This is illustrated by the truth tables, flow charts, and algorithms given below.

Truth Tables For Commutation

Figure 22.1

p	q	(p	V	q)	↔	(q	V	p)
1	1	1	1	1	1	1	1	1
1	1	1	1	0	1	0	1	1
0	1	0	1	1	1	1	1	0
0	0	0	0	0	1	0	0	0

	q	(p	&	q)	↔	(q	&	p)
1	1	1	1	1	1	1	1	1
1	0	1	0	0	1	0	0	1
0	1	0	0	1	1	1	0	0
0	0	0	0	0	1	0	0	0

Flow Charts For Commutation

Figure 22.2

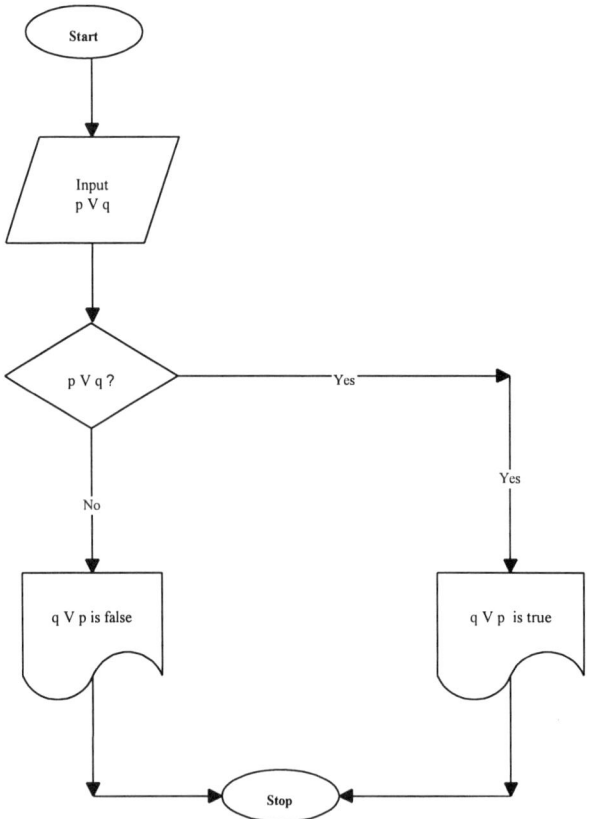

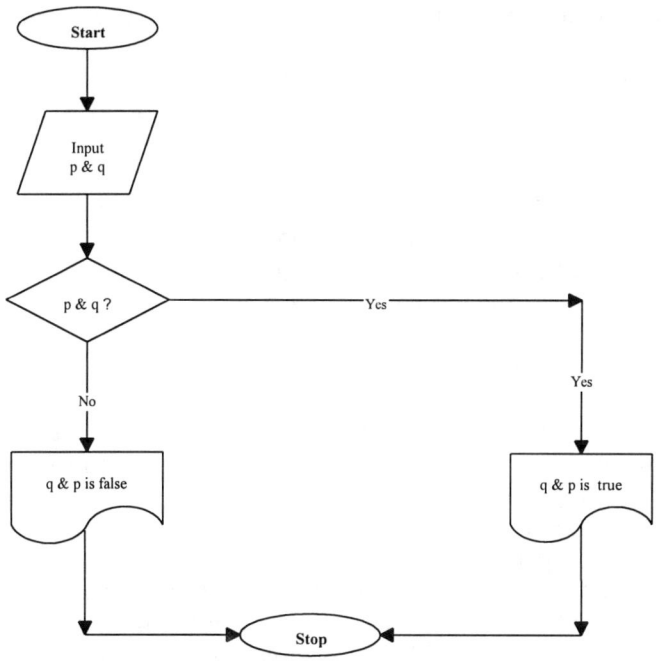

Algorithms For Commutation

Figure 22.3

1. Input value of p V q.
2. If p V q is true,
    (a) q V p is true.
    (b) Stop.
3. q V p is false.
4. Stop.

1. Input value of p & q.
2. If p & q is true,
    (a) q & p is true.
    (b) Stop.
3. q & p is false.
4. Stop.

As with our other principles the logical structure of commutation can be made to **look** more complicated by the addition of variables. Let us take for instance the following problem.

Figure 22.4

1.	{(G V F) V (D & C)} V (B & A)	∴ (A & B) V [(C & D) V (F V G)]
2.	(B & A) V [(G V F) V (D & C)]	1, COM.
3.	(A & B) V [(G VF) V (D & C)]	2, COM.
4.	(A & B) V [(D & C) V G V F)]	3, COM.
5.	(A & B) V [(C & D) V (G V F)]	4, COM.
6.	(A & B) V [(C & D) V (F V G)]	5 COM.

Analysis of the above solution will indicate that the simple logical principle of switching the order of p & q or p V q is the only one that need be used to solve the problem. In the move from line 1 to line 2 [(G V F) V (D & C)] is treated as the p statement while (B & A) is treated as the q statement. In the move from line 2 to line 3, we concentrate on (B & A). B is treated as the p statement. A is treated as the q statement. In the transition from line 3 to line 4 we concentrate on (G V F) V (D & C). (G V F) is treated as the p statement. The move to lines 5 and 6 proceed in the same manner.

The other thing that should be noted is that the solution to the problem proceeds **one** step at a time. This not only simplifies the problem logically but clarifies the documentation of our solution.

PROBLEM SET: COMMUTATION
Demonstrate the validity of the following arguments.

1.
   1.   D & E   ∴ E & D

2.
   1.   D & E   ∴ ~~E & D

3.
   1.   (A & B) → (C & D)   ∴ D&C
   2.   B & A

4.

   1.   [(A & B) V (C & D)] & [(E & F) V (G V H)]

           ∴ [(H V G) V (F & E)] & [(D & C) V (B & A)]

5.

   1.   (A & B) → (C & D)   ∴ (B & A) & (D & ~ ~ C)

   2.   B & A

6.

   1.   A V D   ∴ (B & A) V ((C & D)

   2.   A → B

   3.   D → C

7.

   1.   D → E   ∴ G

   2.   F → (G V H)

   3.   F V D

   4.   ~ (D & E)

   5.   ~ H

8.

   1.   ~[(A & B) V (C & D)] V [(E & F) V (G V H)] ∴ F

   2.   A & B

   3.   J

   4.   J → ~ (G V H)

9.

   1.   J   ∴ ~ U

   2.   K

   3.   ~[(S → T) & (U → V)] → ~ [(K V L) & (J V M)]

   4.   ~ V

10.

   1.   B → C   ∴ ~~[~~(C & B) & A] V [(F & ~~E) & ~~ D]

   2.   A → B

   3.   E → F

   4.   D V A

   5.   D → E

   6.   C → D

1.   Tom and Henry are going to the store. Therefore, Henry and Tom are going to the store.   T, H.

2.   Henry is going to law school or Tom is going to Med. school. Therefore, either Tom is going to Med. school or Henry is going to law school. H, T.

3.   Henry's doing well on the LSAT is a sufficient condition for Henry's going to law school, and Tom's doing well on the MCAT is a necessary condition for Tom's going to Med. school. Therefore, Tom's doing well on the MCAT is a necessary condition for Tom's going to Med. school, and Henry's doing well on the LSAT is a sufficient condition for Henry's going to law school. H, L, T, M.

4.   Henry's doing well on the LSAT is a sufficient condition for Henry's going to law school or Tom's doing well on the MCAT is a necessary condition for Tom's going to Med. school. Therefore, Tom is going to Med. school only if Tom does well on the MCAT or Henry's doing well on the LSAT entails that Henry will go to law school.   H, L, M, T.

5.   If Henry does well on the LSAT, then he is going to law school and if Tom does well on the MCAT then he is going to Med. school. Either Tom will do well on the MCAT or Henry will do well on the LSAT. Therefore, either Henry will go to law school or Tom will go to Med. school.   H, L, T, M.

6.   If production is up and Jones works overtime, then Jones' income will rise and he will buy a new car. Jones will work overtime and production is up. Therefore, Jones will buy a new car and his income will rise.
                           P, O, R, C.

7.   Production is up and Jones will work overtime, or production is down and Jones will be fired; and profits will increase and Jones will be promoted, or profits will slip and Jones will have to move. Therefore, Jones will be promoted and profits will increase, or Jones will have to move and profits will slip; and production is down and Jones will be fired, or production is up and Jones will work overtime.   U, O, D, F, I, P, S, M.

8.   Either the judge is honest and the jury is informed, or it is not the case both that both Jones is innocent and will go free. Jones will go free and is innocent. Therefore, the jury is informed. H, J, I, F.

9. If Jones is innocent, then he will be acquitted. If Jones was in Chicago on Tuesday, then he is innocent. If Jones took flight 202, then he was in Chicago on Tuesday. Therefore, if Jones took flight 202, he will be acquitted and is innocent, and was in Chicago. I, A, C, F.

10. Jones is guilty only if Jones was in Chicago on the first. Jones was in Chicago on the first or Jones was in Miami. Smith was in Chicago on the first and knows Jones well. Smith is either an accomplice of Jones or a victim. If the following is true Jones will be convicted: Smith is either a victim or an accomplice of Jones; Jones being guilty is a sufficient condition for Jones being in Chicago on the first; Smith knows Jones well and was in Chicago on the first; Jones was in Miami or in Chicago on the first. Therefore, Jones will be convicted. G, F, M, S, K, A, V, C.

CHAPTER 23: TAUTOLOGY

$$p \leftrightarrow (p \lor p)$$
$$p \leftrightarrow (p \,\&\, p)$$

The English word tautology (abb. Taut.) comes from two Greek words. "Tautos" and "logos." Tautos means literally "the same." "Logos" means 'word.' Thus, a "tautology" literally means "the same word." A tautology is, therefore, just two different ways of saying the same thing,

There is a sense in which all of the rules of replacement and even all systems of deductive logic are tautologies. In the case of the rules of replacement, we are justified in replacing one phrase with another because, however different they may appear at just a glance, they are logically asserting the same thing. In the case of a valid deductive argument, the conclusion of the argument is already implicitly contained in the premises. Thus, the conclusion strictly speaking does not provide us with any new information but affirms at least a portion of the same material which is already contained in the premises. We might make an analogous point with respect to the use of computers. Computers do not, nor are they designed to, generate new information. They simply present the same information in a different format. Of course, the different arrangement of data may be "new to us" in the sense that we were totally unaware that the results of the program were implicit in the inputs we had given. The arrangement of the same data in different ways is not only frequently useful. It can also sometimes

lead to surprises. We can make still another analogy with the legal
system. When I instruct my attorney with respect to my wishes
regarding my will, I expect him to take the same instructions and
translate them into legally appropriate language. Of course, here the
goal is not the discovery of surprises but their avoidance.

In our present context, tautology is designated as a specific rule of
replacement. What is being asserted is that saying the same thing in
the same way twice is logically equivalent to making that assertion
only once. This is true despite the fact that we have all met people who
seemed sincerely to believe that by repeating the same thing in the
same way they were telling us something new. Repetition may make a
difference in our memory retention or level of boredom but it makes no
logical difference.

Thus, " I have a cat and I have a cat." is equal to "I have a cat."
Analogously, "I either have a dog or I have a dog." is equal to "I have a
dog."

Truth Tables For Tautology

Figure 23.1

P	↔	(p	V	p)
1	1	1	1	1
0	1	0	0	0

Figure 22.2

P	↔	(p	&	p)
1	1	1	1	1
0	1	0	0	0

Figure 23.

Flow Chart For Tautology

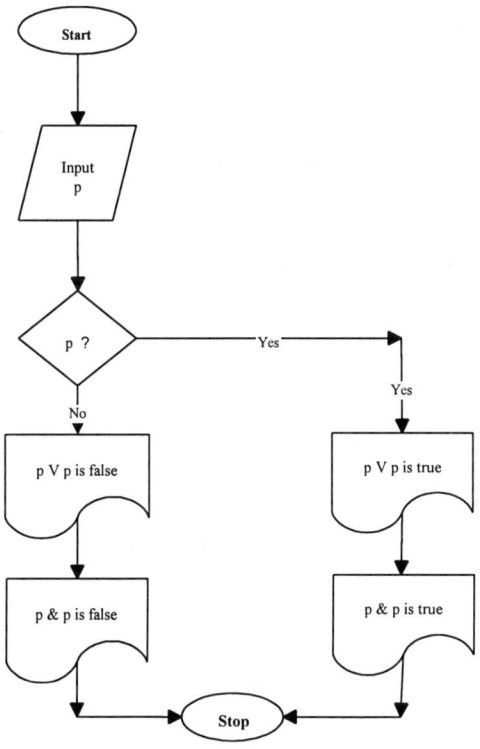

Figure 23.4

Algorithm For Tautology

1. Input value of p.
2. If p is true,
    (a) p V p is true & p & p is true.
    (b) Stop.
3. p V p is false & p & p is false.
4. Stop.

PROBLEM SET: TAUTOLOGY

Demonstrate the validity of the following arguments.

1.
    1.   D   ∴ D & D

2.
    1.   E   ∴ E V E

3.
    1.   D   ∴ (D V D) & (E & E)
    2.   E

4.
    1.   D   ∴ [(D V D) & (E & E)] & [(E & E) & (D V D)]
    2.   E

5.
    1.   [(A & A) V A] V [~ ~ A V (A & A)]   ∴ A

6.
    1.   {(E V E) & E] → (C V ~ D)   ∴ C V X
    2.   D & E

7.
    1.   D → R   ∴ ~ ~ (~ ~ R V X) V (C V R)
    2.   E → C
    3.   (D & D) V (E & E)

8.
    1.   ~ (D & D)   ∴ (C & C) V (F & F)
    2.   (E & E) → (D & D)
    3.   ~ E → C

9.
    1.   {[(E & E) V (E V E)] & E} & {[(D V D) V (D V D)] V D}
                                      ∴ E & D

10.
    1.   (B → C) → A   ∴ G
    2.   (D V E) → A
    3.   (D V E) V (B → C)
    4.   A → G

1.   Jones is working and paying taxes. Therefore, Jones is paying taxes and working, and working and paying taxes. W, P.

2.   Jones is working and paying taxes. Therefore, Jones is rich; or Jones is paying taxes and working, and Jones is working and paying taxes.
W, P, R.

3.   Jones's working is a necessary condition for his paying taxes. Jones is paying taxes and Jones is paying taxes. Therefore, Jones is working.
W, P.

4.   If Jones's paying taxes is a sufficient condition for Jones working or Jones pays taxes only if he is working, then Jones is employed. Jones's working is a necessary condition for Jones's paying taxes. Therefore, Jones is employed. T, W, E.

5.   If the auditor is a partner of his client, he has compromised his independence. If the auditor has substantial financial interest in his client's business, then he has compromised his independence. The auditor is either a partner of his client or has substantial financial interest in his business. Therefore, the auditor has compromised his independence.   P, C, F.

6.   Explanation: Susan is a CPA and is considering doing an external audit for company XYZ.

If the following is false; if Susan is a manager of XYZ then she cannot be independent and Susan is a manager of XYZ only if she cannot be independent, entails that her professional conduct is unethical; then Susan did not refuse the assignment. Susan did refuse the assignment. Susan's being a manager of XYZ is a sufficient condition for her not being able to be independent. Therefore, Susan's conduct is unethical.
M, I, U, R.

7. An increase in revenue is a sufficient condition for an increase in profits and revenue will increase only if profits increase, or if revenue increases then profits will increase; and profits increasing is a necessary condition of revenue increasing or an increase in revenue entails an increase in profits, or an increase in revenue necessarily leads to an increase in profits. Revenue will increase. Therefore, profits will increase. R, P.

8. Jones is guilty and rich. Therefore, Jones is rich and guilty, or guilty and rich; and Jones is rich and guilty, or guilty or rich. G, R.

9. It is not the case that, Smith is rich or it is not the case that he is not rich. If Smith is rich only if he is a successful businessman, then Smith is rich. Smith's owning a large house, is a sufficient condition for Smith's being a successful businessman being a necessary condition for his being rich. Therefore, Smith does not own a large house. R, B, H.

10. If Smith is associated with organized crime, then he tested positive only if he was taking illegal drugs. If Smith did not associate with organized crime, then he tested positive only if he was on prescribed medication. Smith either associated with organized crime or he did not associate with organized crime but he is a first rate attorney anyway. If Smith's testing positive for drugs is a sufficient condition for his taking illegal drugs, then he must be investigated further. If Smith's testing positive for drugs entails that he is on prescribed medication, then he must be investigated further. Therefore, Smith must be investigated further.

    O, P, D, M, A, I.

CHAPTER 24: ASSOCIATION

$$[p \lor (q \lor r)] \leftrightarrow [(p \lor q) \lor r]$$
$$[p \ \& \ (q \ \& \ r)] \leftrightarrow [(p \ \& \ q) \ \& \ r]$$

The rule of replacement known as association (abb. Assoc.) like a number of the other rules we have studied is intuitively obvious upon a little reflection. One way of putting it is to say that it simply involves a change in punctuation so that a given variable becomes more closely grouped with one variable rather than another. Consider the following statements:

I have a cat, and also have a dog and a horse.
I have a cat and a dog, and also have a horse.

These two statements are obviously logically identical although in ordinary English the change in the placement of the comma may serve to emphasize the relationship to one variable more than the relationship to the others. As we have indicated earlier, symbolic logic is capable of much greater precision than ordinary English. There is, however, frequently a loss of the nuances of meaning. Take, for example, the following statements in ordinary English.

I have a brother, and a sister and a cat. B, S,C.
I have a brother and a sister, and a cat.

The two statements in ordinary English may connate a different meaning. For instance the first statement might indicate that I regard my sister and my cat of equal importance. In symbolic logic, however, the meaning of B & (S & C) is identical to the meaning of (B & S) & C.

There are two kinds of association -- the "inclusive or" and the "conjunction" as illustrated by the truth tables, flow charts and algorithms given below.

Figure 24.1

Truth Table For Association (Inclusive Or)

p	q	r	[p	V	(q	V	r)]	↔	[(p	V	q)	V	r]
1	1	1	1	1	1	1	1	1	1	1	1	1	1
1	1	0	1	1	1	1	0	1	1	1	1	1	0
1	0	1	1	1	1	1	1	1	1	1	0	1	1
1	0	0	1	1	0	0	0	1	1	1	0	1	0
0	1	1	0	1	1	1	1	1	0	1	1	1	1
0	1	0	0	1	1	1	0	1	0	1	1	1	0
0	0	1	0	1	0	1	1	1	0	0	0	1	1
0	0	0	0	0	0	0	0	1	0	0	0	0	0

Figure 24.2

Flow Chart For Association (Inclusive Or)

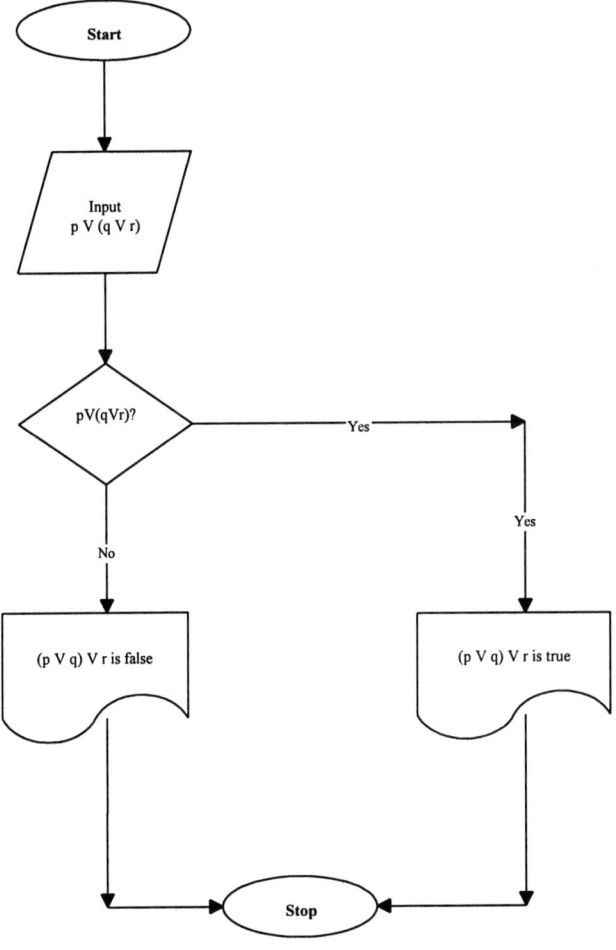

Figure 24.3

Algorithm For Association (Inclusive Or)

1.   Input value of [p V (q V r)]
2.   If [p V (q V r)] is true,
         (a) [(p V q) V r] is true.
         (b) Stop.
3.   [(p V q) V r] is false.
4.   Stop.

Because of the logic of the 'inclusive or', if **any** of the variables p, q, or r is true, then both [p V (q V r)] & [(p V q) V r] are true. The flow chart and algorithm given below illustrate this relationship as does the truth table which shows the whole phrase true when only one of the variables p, q, or r is true.

Figure 24.4

Truth Table For Association When Only One Of The Variables Is True

p	q	r	[p	V	(q	V	r)]	↔	[(p	V	q)	V	r]
1	0	0	1	1	0	0	0	1	1	1	0	1	0
0	1	0	0	1	1	1	0	1	0	1	1	1	0
0	0	1	0	1	0	1	1	1	0	0	0	1	1

Figure 24.5

Flow Chart For Association (Inclusive Or)

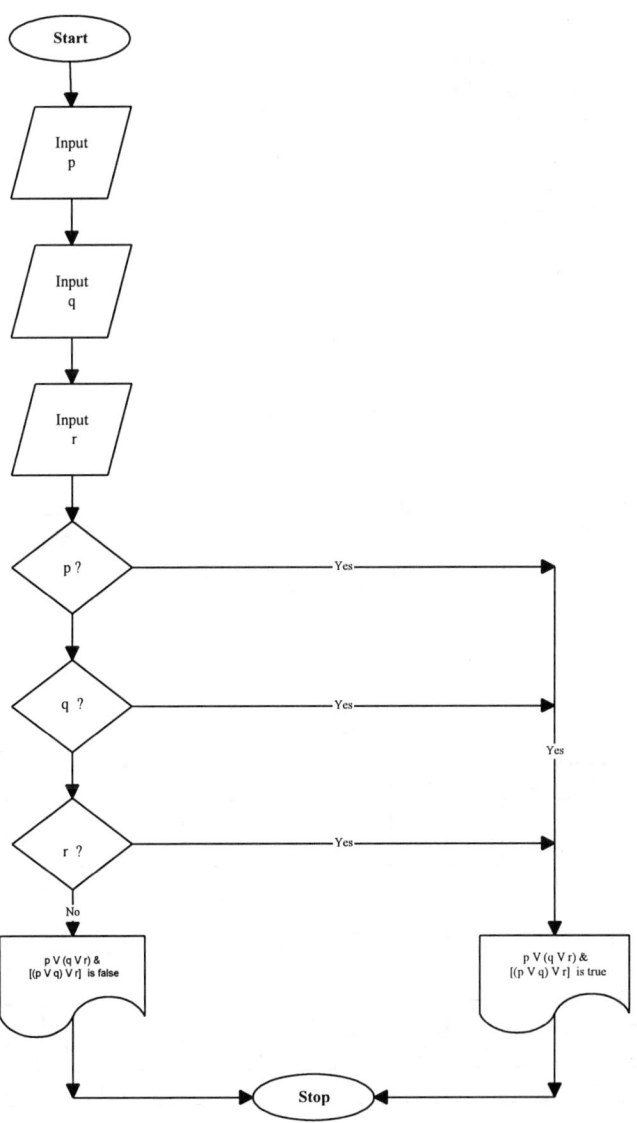

Figure 24.6

Algorithm For Association (Inclusive Or)

1.   Input value of p.
2.   Input value of q.
3.   Input value of r.
4.   If p is true,
         (a) [p V (q V r)] & [(p V q) V r] is true.
         (b) Stop.
5.   If q is true,
         (a) [p V (q V r)] & [(p V q) V r] is true.
         (b) Stop.
6.   If r is true,
         (a) [p V (q V r)] & [(p V q) V r] is true.
         (b) Stop.
7.   [p V (q V r)] & [(p V q) V r] is false.
8.   Stop.

In one respect truth conditions for inclusive or statements are the opposite of truth conditions for a conjunction. For inclusive or statements to be true **only one** variable need be true. For conjunction statements to be true **all** the variables must be true.

Figure 24.7

Truth Table For Association (Conjunction)

p	q	r	[p	&	(q	&	r)]	↔	[(p	&	q)	&	r]
1	1	1	1	1	1	1	1	1	1	1	1	1	1
1	1	0	1	0	1	0	0	1	1	1	1	0	0
1	0	1	1	0	0	0	1	1	1	0	0	0	1
1	0	0	1	0	0	0	0	1	1	0	0	0	0
0	1	1	0	0	1	1	1	1	0	0	1	0	1
0	1	0	0	0	1	0	0	1	0	0	1	0	0
0	0	1	0	0	0	0	1	1	0	0	0	0	1
0	0	0	0	0	0	0	0	1	0	0	0	0	0

Figure 24.8

Flow Chart For Association (Conjunction) #1

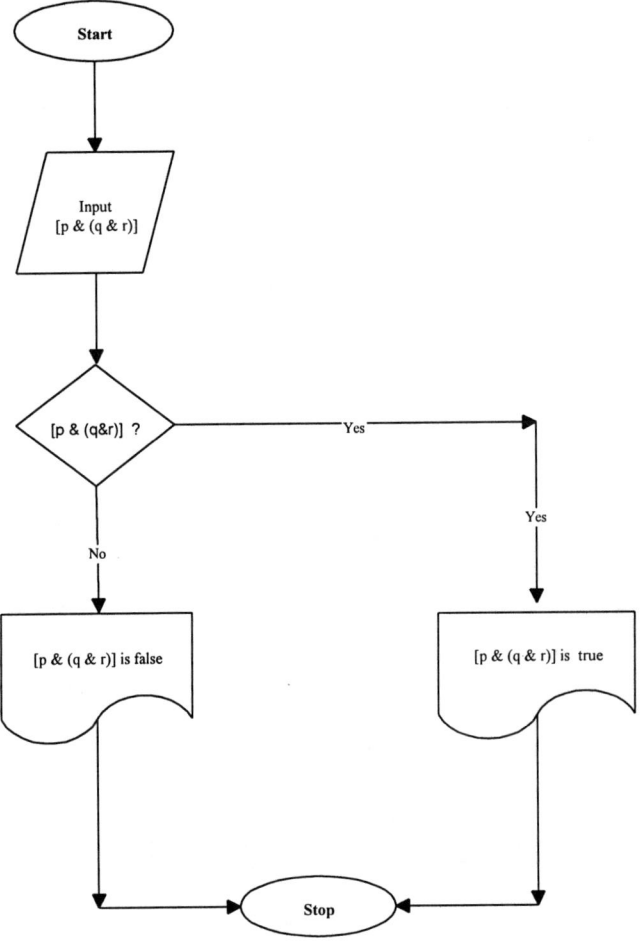

Figure 24.9

Flow Chart For Association (Conjunction) #2

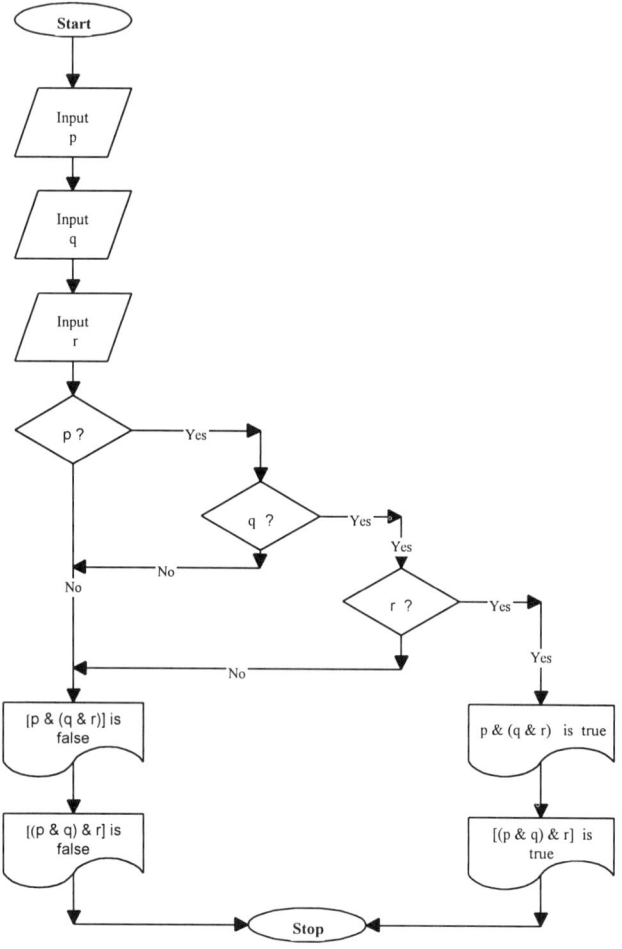

Figure 24.10

Algorithm For Association (Conjunction) #1

1. Input value of [p & (q & r)].
2. If p & (q & r) is true,
    (a) [p & (q & r)] is true.
    (b) Stop.
3. [p & (q & r)] is false.
4. Stop.

Figure 24.11

Algorithm For Association (Conjunction) #2

1. Input value of p.
2. Input value of q.
3. Input value of r.
4. If p is false,
    (a) [p & (q & r)] & [(p & q) & r)] is false.
    (b) Stop.
5. If q is false,
    (a) [p & (q & r)] & [(p & q) & r)] is false.
    (b) Stop.
6. If r is false,
    (a) [p & (q & r)] & [(p & q) & r)] is false.
    (b) Stop.
7. [p & (q & r)] & [(p & q) & r)] is true.
8. Stop.

Below are problems which require the use of the principle of association in their solutions. Some procedural rules may be helpful. In the first place, the principle may be applied only to variables which are adjacent to one another. Take for example the transition of p & (r & q) to (p & q) & r. This requires two steps as shown below.

Figure 24.12

1.	p & (r & q)		∴ (p & q) & r
2.	p & (q & r)	1, COM.	
3.	(p & q) & r	2, ASSOC.	

The above satisfies correct procedure. The solution given below in 24.13 skips a step and is incorrect.

Figure 24.13

1.	p & (r & q)		∴ (p & q) & r
2.	(p & q) & r	1, ASSOC.	

Secondly, the principle of association really involves two logical operations -- a disassociation of two variables from each other and then the association of one variable with another. For example, in the transition from p & (q & r) to (p & q) & r, we first disassociate the q from the r and then associate the q with the p. This has implications for the following. The first solution given below in figure 24.14 is correct. The second given in 24.15 is incorrect.

Figure 24.14

1.	(r & p) & (q & s)		∴ (p & q) & (r & s)
2.	[(r & p) & q] & s	1, ASSOC.	
3.	[r & (p & q)] & s	2, ASSOC.	
4.	[(p & q) & r] & s	3, COM.	
5.	(p & q) & (r & s)	4, ASSOC.	

Figure 24.16

1.	(r & p) & (q & s)		∴ (p & q) & (r & s)
2.	[r & (p & q)] & s	1, ASSOC.	
3.	[(p & q) & r] & s	2, COM.	
4.	(p & q) & (r & s)	3, ASSOC.	

Why is the second solution incorrect? Strictly speaking, it makes a documentation error rather than a logical error. In step number two, we disassociate the p from the r statement and also disassociate the q from the s statement before associating the p and q statements. Thus, rather than one disassociation and one association, we have two disassociations and one association in one step. The first solution avoids this documentation mistake.

Thirdly, and most importantly, association works when three variables are linked by conjunction, or it works when three variables are linked by the inclusive or. It is not appropriate when three variable are linked by a combination of conjunction and inclusive or. For example, the principle of association cannot be applied to regroup p & (q V r) so that it becomes (p & q) V r. This can be shown by an examination of the truth table given in figure 24.16.

Figure 24.16.

p	q	r	p	&	(q	V	r)	↔	(p	&	q)	V	r
1	1	1	1	1	1	1	1	1	1	1	1	1	1
1	1	0	1	1	1	1	0	1	1	1	1	1	0
1	0	1	1	1	0	1	1	1	1	0	0	1	1
1	0	0	1	0	0	0	0	1	1	0	0	0	0
0	1	1	0	0	1	1	1	0	0	0	1	1	1
0	1	0	0	0	1	1	0	1	0	0	1	0	0
0	0	1	0	0	0	1	1	0	0	0	0	1	1
0	0	0	0	0	0	0	0	1	0	0	0	0	0

In the logical possibilities depicted in rows 5 and 7 the statement variables are true and false under different conditions thus showing that they are not bi-conditionals.

PROBLEM SET: ASSOCIATION
Demonstrate the validity of the following arguments

1.
   1. D & (A & B)   ∴ (D & A) & B

2.
   1. D & (A & B)   ∴ B

3.
    1.   (C V B) V A   ∴ (A V B) V C

4.
    1.   (C & B)   ∴ (A & B) & C
    2.   A

5.
    1.   (A V D) V (B V C)   ∴ (A V B) V (C V D)

6.
    1.   (A & D) & (B & C)   ∴ (A & B) & (C & D)

7.
    1.   (C V B)   ∴ (A V B) V C

8.
    1.   D V (B V C)   ∴ (E V F) V G
    2.   D → F
    3.   B → (G V E)
    4.   ~ C

9.
    1.   E → (G V J)   ∴ (G V H) V (I V J)
    2.   F V E
    3.   F → (I V H)

10.
    1.   A → B   ∴ A → [(A & D) & (C & B)]
    2.   B → C
    3.   C → D

1. Lisa is majoring in Spanish and is going to Madrid, and also majoring in psychology. Therefore, Lisa is going to Madrid, and majoring in psychology and Spanish. S, M, P.

2. Heidi is majoring in Spanish or minoring in business, or going to England. Therefore, Heidi is going to England or majoring in Spanish, or minoring in business. S, B, E.

3. South has been indicted. Therefore, South will be convicted or he will be set free, or he has been indicted; and South will be convicted, or he will be set free or he has been indicted. I, C, F.

4. South has been indicted, or he will go to jail or he will be placed on probation. South will not go to jail. Therefore, South has been indicted or he will be placed on probation. I, J, P.

5. Susan was caught with possession of drugs, but was both at the scene of the robbery and home one half hour later. If Susan was caught with possession of drugs and was at the scene of the robbery, then she will be convicted. Therefore, Susan will be convicted. D, R, H, C.

6. Jones is a hedonist or a Kantian. Therefore, Jones is a Kantian or a Platonist, or Jones is a hedonist. H, K, P.

7. If Rawls is right, then either equality or utility is primary. If MacIntryre is right, then either virtue is primary or rules are primary. Either MacIntryre or Rawls is right. Therefore, either utility is primary or virtue is primary, or equality is primary or rules are primary. R, E, U, M, V, P.

8. If Rawls is right, then equality is primary. If equality is primary, then utilitarianism is inadequate. If utilitarianism is inadequate, then Bentham is wrong. Therefore, if Rawls is right; then Bentham is wrong and Rawls is right, and utilitarianism is inadequate and equality is primary. R, E, U, B.

9. If production goes up; then second shift will work overtime and second shift will be paid overtime pay, or we will add a third shift and third shift will be paid more than second shift. If production goes down, then no one will work overtime. Production will go up or down. Therefore, no one will work overtime, or we will add a third shift and third shift will be paid more than second shift; or second shift will be paid overtime pay and second shift will work overtime. P, S, O, T, M, D, N.

10. If Bennett is brash and Bennett is a philosopher, and he is politically astute, then Bennett is the new drug Czar. Bennett is a philosopher and politically astute. Bennett is brash. Therefore, Bennett is the new drug Czar. B, P, A, C.

CHAPTER 25: TRANSPOSITION

$$(p \to q) \leftrightarrow (\sim q \to \sim p)$$

The name for transposition (abb. Trans.) comes from the English word transpose which means to change the relative position of two variables. The rules for the application of transposition are two. (1) Change the relative position of both variables. Thus, the p and the q statement change places. (2) Change the designation of each variable from positive to negative or from negative to positive, adding a sign if necessary or subtracting when possible. Consider the examples given below.

Figure 25.1

(1) $(A \to B) \leftrightarrow (\sim B \to \sim A)$
(2) $(\sim C \to D) \leftrightarrow (\sim D \to C)$

The first example shows a simple reversal of A and B and the change to the negation of both the A and the B. In the second example, since we are able to change the value of the C statement by subtracting the negation sign we do so. In order to change the D statement we must add a negation sign as is shown in the example.

Figure 25.2

How about $\sim\sim\sim C \to \sim\sim D$?
This would become $\sim D \to \sim\sim C$.

Logic of the Computing Sciences

As with our other principles, a truth table, a flow chart and an algorithm are given below.

Figure 25.2

Truth Table For Transposition

p	q	(p	→	q)	↔	(~	q	→	~	p)
1	1	1	**1**	1	1	0		**1**	0	
1	0	1	**0**	0	1	1		**0**	0	
0	1	0	**1**	1	1	0		**1**	1	
0	0	0	**1**	1	1	1		**1**	1	

Figure 25.4

Flow Chart for Transposition

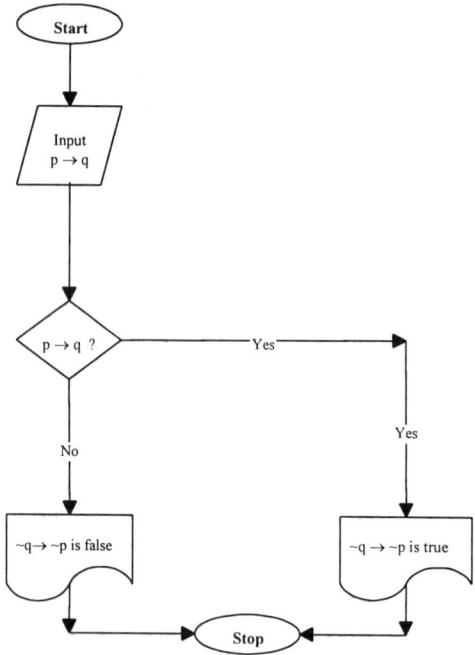

Figure 25.5

Algorithm For Transposition

1. Input value of p → q.
2. If p → q is true,
   (a) ~q → ~p is true.
   (b) Stop.
3. ~q → ~p is false.
4. Stop.

PROBLEM SET: TRANSPOSITION
Demonstrate the validity of the following arguments.

1.
   1. D → C   ∴ ~C → ~D

2.
   1. ~D → ~C   ∴ C → D

3.
   1. ~~C → D   ∴ ~D → ~C

4.
   1. ~C → ~B   ∴ A → C
   2. ~B → ~A

5.
   1. ~A → ~B   ∴ A V C
   2. ~C → D
   3. B V ~D

6.
   1. ~[~(~H → ~G) → ~(~F → ~E)] → ~[~(~D → ~C) → ~(~B → ~A)]
      ∴ [(A → B) → (C → D)] → [(E → F) → (G → H)]

7.
   1. ~(~B → ~A) → ~(~D → ~C)   ∴ B
   2. C → D
   3. A

8.
    1. C V F   ∴ (A V E) V (B V D)
    2. ~(D V E) → ~F
    3. ~(A V B) → ~C

9.
    1. ~B → ~A   ∴ (B & A) V (D & C)
    2. ~D → ~C
    3. C V A

10.
    1. C   ∴ A
    2. ~A → B
    3. ~C V ~B

1.   If Jane goes to law school, then she will be a trial lawyer. Therefore, if Jane doesn't become a trial lawyer, she will not go to law school. L, T.

2.   If the primary reason for the South's losing the war was not Jefferson Davis's leadership then economic deficiency is the best explanation. Therefore, if economic deficiency is not the best explanation, then the primary reason for the South's losing the war was Jefferson Davis's leadership. J, D.

3. If it is not the case that the following do not explain the defeat of the confederacy, then Jefferson Davis's lack of leadership is the explanation: the Economic superiority of the North, the failure of England and France to intervene, an excessive emphasis on democracy in the South, and superior Northern military strategy. Therefore, if Jefferson Davis's lack of leadership is not the explanation, then the following is the explanation: the economic superiority of the North, the failure of England and France to intervene, an excessive emphasis on democracy in the South and superior Northern military strategy. L, E, F, D, M.

   Let L equal "the explanation of the defeat of the confederacy is Jefferson Davis's lack of leadership."
   Let E equal "explanation of the defeat of the confederacy is the Economic superiority of the North."
   Let F equal "the explanation of the defeat of the Confederacy is the failure of England and France to intervene."
   Let D equal "the explanation of the defeat of the Confederacy is an excessive emphasis on democracy."
   Let M equal "the explanation of the defeat of the Confederacy is superior Northern military strategy."

4. The following entails that some national exams are important: Arthur is going to law school only if he does well on the LSAT and Barbara's doing well on the MCAT is a sufficient condition for Barbara's going to Med. school. If Arthur does not do well on the LSAT, then Arthur is not going to law school. If Barbara does not go to Med. school, then she will not do well on the MCAT. Therefore, some national exams are important.

$$N, A, L, M, B, N.$$

5. If Bentham is right entails that pleasure is the good, then hedonism is a correct doctrine. Therefore, if hedonism is not a correct doctrine, then its not true that if pleasure is not the good then Bentham is not right.

$$B, P, H.$$

6. If pleasure is the good, then Bentham is right. If obedience to a universal rational law is the good, then Kant is right. Either Kant is not right or Bentham is not right. Therefore, either pleasure is not the good or obedience to a universal rational law is not the good.   P, B. O. K.

7. If Whitehead is right and Augustine is right, then God is both temporal and non temporal. Therefore, it is not the case that God is both temporal and non temporal only if it isn't true that both Whitehead and Augustine are right. W, A, T, N.

8. If Susan does not love James with an agape type love, then she does not have a self giving love for him. If Susan does not have a self giving love for James, then she does not have an agape type love for him. Therefore, if Susan loves James with a self giving love, then she loves James with an agape type love; and if Susan loves James with an agape type love, then she love James with a self giving love. A, G.

9. If Whitehead is right then God is involved in time, and if Brightman is right then God is finite; entails that if process philosophers are correct, then God is not non temporal and he is not omnipotent. If God is not finite, then Brightman is not right. God is not involved in time only if Whitehead is not right. Therefore, it is not the case both that God is not omnipotent and God is not non temporal only if process philosophers are not correct. W, T, B, F, P, N, O.

10. James should marry Susan only if he has a self giving love for her. James does not have a self giving love for Susan, if he does not have an agape type love for her. Therefore, if James does not have an agape type love for her, he should not marry Susan. M, G, A.

CHAPTER 26:   MATERIAL IMPLICATION
(p → q) ↔ (~ p V q)

Reflection on the meaning of implication (abb. Impl.) makes clear that (p → q) is a bi-conditional of (~ p V q) although this may come as a surprise to some when it is first encountered. (p → q) asserts that if we have a p, then we will necessarily have a q. Now either we must have a p or a ~ p. If we have a p, then we have a q, in which case (~ p V q) is true, since the "inclusive or" asserts merely that **at least one** of the disjuncts is true. If we do not have a p, then we have ~ p, in which case (~ p V q) is true.

Rules for changing an entailment to an inclusive or statement or vice a versa can be readily summarized (1) change the entailment (→) to an or (V) or an (V) to an entailment. (2) Change the sign in front of the p statement by adding or subtracting a negation sign. Thus, p → q becomes ~p V q. ~~~p → q would become ~~p V q.

Figure 26.1

Truth Table For Material Implication

p	q	(p	→	q)	↔	(~		p	V q)
1	1	1	1	1	1	0		1	1
1	0	1	0	0	1	0		0	0
0	1	0	1	1	1	1		1	1
0	0	0	1	0	1	1		1	0

Flow Charts For Material Implication

Figure 26.2

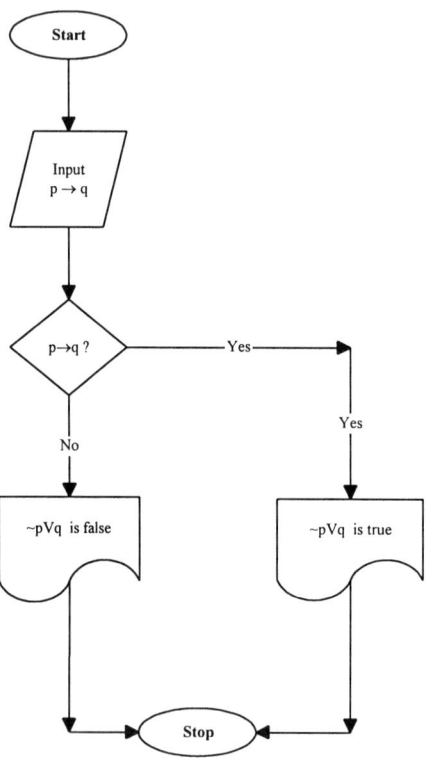

Figure 26.3

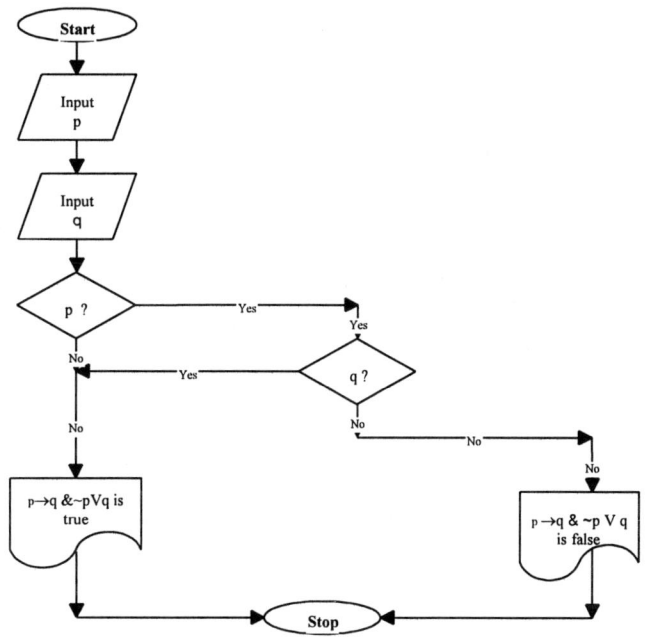

Algorithms For Material Implication

Figure 26.4

1.   Input value of p → q.
2.   If p → q is true,
         (a) ~p V q is true.
         (b) Stop.
3.   ~p V q is false.
4.   Stop.

Figure 26.5

1.   Input value of p.
2.   Input value of q.
3.   If p is true,
         (a) if q is false, p → q is false & ~p V q is false.
         (b) Stop.
4.   p → q is true & ~p V q is true.
5.   Stop.

PROBLEM SET: MATERIAL IMPLICATION
Demonstrate the validity of the following arguments.

1.
      1.   X → Y   ∴ ~X V Y
2.
      1.   ~A V B   ∴ A → B
3.
      1.   ~A → B   ∴ A V B
4.
      1.   ~A → (~B → C)   ∴ (A V B) V C
5.
      1.   A V B   ∴ ~A → (~A & B)
6.
      1.   ~A V B   ∴ (A & B) V (X & Y)
      2.   ~X V Y
      3.   X V A
7.
      1.   ~D → ~C   ∴ A → D
      2.   ~B → ~A
      3.   ~~C V ~B
8.
      1.   ~C V D   ∴ ~A V [(A & B) & (C & D)]
      2.   B V ~A
      3.   ~B V C
9.
      1.   B V ~A   ∴ ~E → (A → D)
      2.   ~(D V E) → ~C
      3.   ~B V C

10.

1. $H \rightarrow I$ $\therefore J \rightarrow (\sim I \rightarrow \sim F)$
2. $F \rightarrow G$
3. $G \rightarrow H$

1. If Marx is right, then heresy in the Middle Ages was a reflection of class conflict. Therefore, either Marx is not right or heresy in the Middle Ages was a reflection of class conflict. M, C.

2. If Marx is right, then the causes of the Reformation are not theological. Therefore, either the causes of the Reformation are not theological or Marx is not right. M, R.

3. Marx is not right or the laboring class will become more oppressed. Therefore, the laboring class will become more oppressed, if Marx is right. M, L.

4. If Marx is not right, then the labor theory of value is not right. Therefore, either Marx is right or the labor theory of value is not right. M, L.

5. If it is not the case that ancient Israel's having strict dietary laws entails that physical hygiene was the chief concern, then ritual cleanliness was the chief concern. Therefore, either physical hygiene being a chief concern was a necessary condition for ancient Israel's having strict dietary laws, or ritual cleanliness was the chief concern. D, H, R.

6. If it is not the case that ancient Israel's having strict dietary laws entails that physical hygiene was the chief concern, then ritual cleanliness was the chief concern. Physical hygiene was not the chief concern. Therefore, ancient Israel had strict dietary laws only because ritual cleanliness was the chief concern. D, H, R.

7. The fact that the Hebrews prohibited adultery entails that God forbade it, or the fact that the Hebrews prohibited adultery is only because adultery violated male property rights. Therefore, the Hebrews did not prohibit adultery, or adultery violated male property rights or God forbade adultery. H, G, M.

8. If Moses supposes that his toeses are roses, then Moses supposes erroneously. Therefore, George Washington being president entails that Moses supposes erroneously, or Moses does not suppose that his toeses are roses. R, E, W.

9. Either Rawls is not right or equality is the chief criterion of justice, and Bentham is not right or utility is the chief criterion of justice. If Rawls is not right then Bentham is right. Therefore, either utility is the chief criterion of justice or equality is the chief criterion of justice.
R, E, B, U.

10. Either Rawls is not right or equality is the primary criterion of justice. The difference principle is not sound, or inequality is justified only if it benefits the least advantaged. It is not the case that inequality being justified entails that it benefits the least advantaged, or the social structure of capitalism is partially justified. Either equality is not the primary criterion of justice or the difference principle is sound. Therefore, if Rawls is right the social structure of capitalism is partially justified. R, E, D, I, B, C.

CHAPTER 27: EXPORTATION
[(p & q) → r] ↔ [p → (q → r)]

The principle of exportation (abb. Exp.) affirms that if two variables (p & q) entail another (r), then the first variable (p) entails that the second variable (q) entails (r). Perhaps using a concrete example may help fix this in the mind. If having a car and fuel entails that I can get to Atlanta (C & F) → A, then we may safely say that if I have a car, then if I have fuel I can get to Atlanta C → (F → A). Or put another way, if the first condition (having a car) is met **then** the meeting of the second condition (having fuel) will be sufficient to insure that I can get to Atlanta.

Neither C nor F by itself, however, is sufficient to insure getting to Atlanta. That is to say we are not affirming (C → A), nor are we affirming (F → A). What we are affirming is that having a car is sufficient to guarantee that, if I have fuel I can get to Atlanta.

Logic of the Computing Sciences

Figure 27.1

Truth Table For Exportation

p	q	r	[(p	&	q)	→	r]	↔	[p	→	(q	→	r)]
1	1	1	1	1	1	1	1	1	1	1	1	1	1
1	1	0	1	1	1	0	0	1	1	0	1	0	0
1	0	1	1	0	0	1	1	1	1	1	0	1	1
1	0	0	1	0	0	1	0	1	1	1	0	1	0
0	1	1	0	0	1	1	1	1	0	1	1	1	1
0	1	0	0	0	1	1	0	1	0	1	1	0	0
0	0	1	0	0	0	1	1	1	0	1	0	1	1
0	0	0	0	0	0	1	0	1	0	1	0	1	0

Flow Charts For Exportation

Figure 27.2

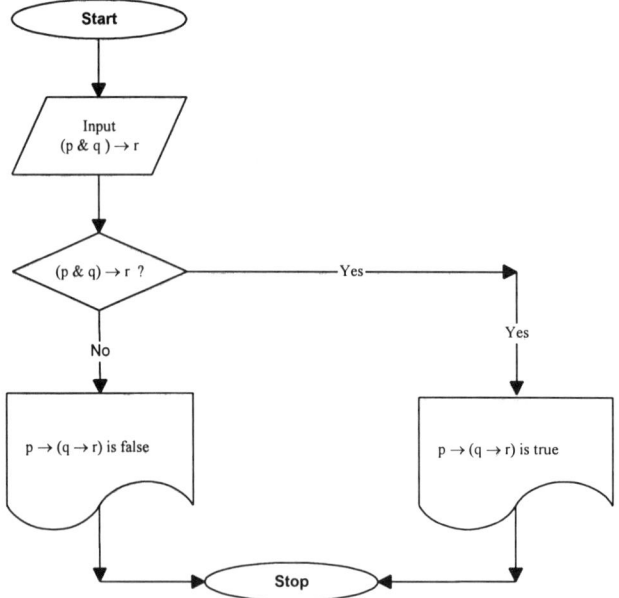

Figure 27.3

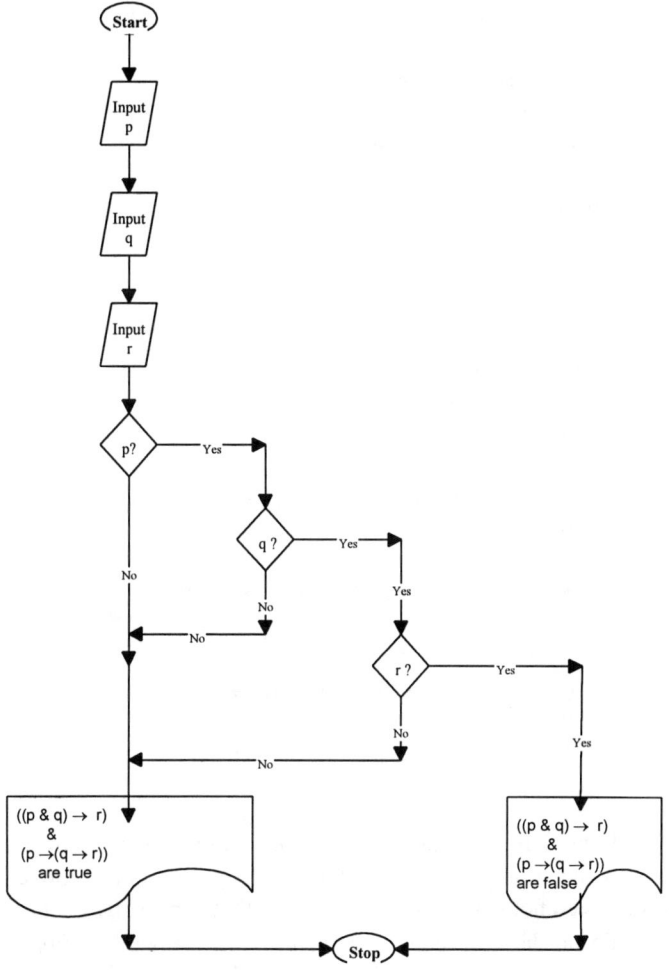

Algorithms For Exportation

Figure 27.4

1.   Input value (p & q) → r.
2.   If (p & q) → r is true,
         (a) p → (q → r) is true.
         (b) Stop.
3.   p → (q → r) is false.
4.   Stop.

Figure 27.5

1.   Input value of p.
2.   Input value of q.
3.   Input value of r.
4.   If p is false,
         (a) (p & q) → r is true & p → (q → r) is true.
         (b) Stop.
5.   If q is false,
         (a) (p & q) → r is true & p → (q → r) is true.
         (b) Stop.
6.   If r is true,
         (a) (p & q) → r is true & p → (q → r) is true.
         (b) Stop.
7.   (p & q) → r is false & p → (q → r) is false.
8.   Stop.

The flow charts and algorithms given above can be shown to be correct by the application of truth tables. Insight into their correctness, however, need not rest on truth tables. All that is necessary is to reflect once again on the truth conditions of p → q. Remember, first of all, the principle that a false p statement will always make an entailment true. Remember secondly that a true q statement will always make an entailment true. Thus, if the A statement of (A & B) → C is false then (A & B) which is the entire p statement will be false regardless of the value of B the statement. Thus, the p statement will be false making the entire statement true (A & B) → C.

0 0   1

When we do an exportation the p statement is still false and the entire statement is true. A → (B → C)

0 1

The result is the same if the A statement is true and B statement is false. (A & B) → C

1 0 0 1

When we do an exportation both the first and second principles come into play since our p statement is true but the false B makes our q statement true and thus the whole statement true. A → (B → C).

1 1 0 1

Of course if A, B and C are all true then our statement is true in any case. As is shown in figure 27.6

Figure 27.6

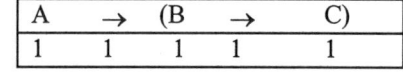

The only condition making for a false statement requires a true A, a true B and a false C as is shown in figure 27.7.

Figure 27.7

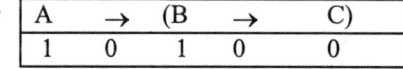

PROBLEM SET: EXPORTATION

Demonstrate the validity of the following arguments.

1.

    1. D → (E → F)   ∴ (D & E) → F

2.

    1. [(D & E) → F] → Z   ∴ Z

    2. E → (D → F)

3.

    1. ~ F   ∴ ~ (D & E)

    2. E → (D → F)

4.

    1. (A & B) → C   ∴ B → C

    2. F → A

    3. F

5.

    1. (A & B) → [(C & D) → (E & F)]

        ∴   A → { B → {C → [D→ (E & F)]}}

6.

    1. (H & I) → J   ∴ (J V ~ H) V ~ I

7.

    1. A V (B V C)   ∴ (~ A & ~ B) → C

8.

    1. A V (C V D)   ∴ [~ I & (H & F)] → G

    2. ~ A V (~ F V G)

    3. (~ C → D) → (~ H V I)

9.

    1. [A → (B → C)] & {[D → (E → F)]} → G

          ∴ {[(A & B) → C] & [(D & E) → F]} → G

10.

    1. [(~ A V ~B) V C] → {[(~ B V ~ F) V G] → (~ I V H)}

        ∴ {[(A & B) → C] & [(B & F) → G]} → (I → H)

1.   If the college is meeting all basic criteria and if it effectively represents itself to the accrediting association, then the college will be re-accredited. Therefore, if the college is meeting all basic criteria, then effectively representing itself to the accrediting association is a sufficient condition for the college being re-accredited. B, E, R.

2.   If statistical data is complete and accurate, then if it is sufficiently precise it will be helpful in decision making. Therefore, if statistical data is sufficiently precise and accurate, and complete, then it will be helpful in decision making. C, A, P, H.

3. If Captain Wirz obeyed orders and law has nothing to do with morality, then Wirz should be exonerated from the atrocities committed at the prisoner of war camp in Andersonville, Georgia. Therefore, if it is not true that Captain Wirz's having obeyed orders entails that Wirz should be exonerated from the atrocities committed at the prisoner of war camp in Andersonville, Georgia, then it is not true that law has nothing to do with morality. O, L, E.

4. Fixed costs are not high, or if variable costs are low then revenue is greater than variable costs. Therefore, if fixed costs are high and variable costs are low, then revenue is greater than variable costs.
F, V, R.

5. Fixed cost are low or variable costs are high, or increased revenue from each additional unit will be greater than variable costs. Therefore, if fixed costs are not low and variable costs are not high, then increased revenue from each additional unit will be greater than variable costs.
F, V, R.

6. Soap Opera Problem: Either its not true that Mary is either not seeing Jack or she is seeing Henry; or the following is true its false that Donald is not seeing Eleonor or he is seeing Patty, or else Donald is really married to Suzy. Therefore, if Mary's seeing Jack entails that she is seeing Henry and Donald's seeing Eleonor entails he is seeing Patty, then Donald is really married to Suzy. J, H, E, P, S

7. It is not true that the following is a sufficient condition for Jim becoming wealthy: if fixed costs are high, then if variable costs are low then increased revenue from every additional unit will be greater than variable costs. Jim will not become wealthy. Therefore, if revenue from every additional unit will not be greater than variable costs then the following are not both true: fixed costs are high and variable costs are low.
W, F, V, R.

8. Soap Opera Problem: The following is a sufficient condition for guaranteeing that Margaret will go to the Bahamas only if she marries Sam: Sam will get a divorce, and Jane's having an affair with Sam entails that Jane will divorce Bill. Either Jane will divorce Bill or Jane is not having an affair with Sam. Therefore, if Sam gets a divorce and Margaret goes to the Bahamas, then Margaret will marry Sam.

   M = Margaret will go to the Bahamas
   S = Margaret will marry Sam
   D = Sam will get a divorce
   J = Jane is having an affair with Sam
   B = Jane will divorce Bill

9. Either its not true both that behaviorism is not true and Skinner is not right, or man is either free or determined. Therefore, if behaviorism is not true and Skinner is not right and if in addition man is not determined, then man is free.   B, S, F, D.

10. If God is good and God is omnipotent, then man is either free or there is not evil in the world. There is evil in the world. Therefore, the following are not all true: God is good and God is omnipotent, and man is not free.
   G = God is good
   F = Man is free
   O = God is omnipotent
   E = There is evil in the world.

CHAPTER 28: MATERIAL EQUIVALENCE

$$(p \leftrightarrow q) \leftrightarrow [(p \rightarrow q) \& (q \rightarrow p)]$$
$$(p \leftrightarrow q) \leftrightarrow [(p \& q) \lor (\sim p \& \sim q)]$$

Material Equivalence (abb. Equiv.) can be understood by considering its meaning in a number of different ways. One way is to see it as a definition of what we mean by bi-conditional. Two statements are bi conditionals if they are true and false under the same conditions. This is what the second explanation listed above asserts. Either p and q are both true in which case (p & q) would be true or they are both false, in which case ($\sim$ p & $\sim$ q) would be true. In the first version cited above we are told that if (p $\leftrightarrow$ q), then both (p $\rightarrow$ q) is true **and** (q $\rightarrow$ p) is true. If you do a truth table on **(p $\rightarrow$ q) and**

(q $\rightarrow$ p) you will find that the only way they can be simultaneously true or false is if **both** p and q are true (p & q) or **both** p and q are false ($\sim$ p & $\sim$ q).

Material Equivalence can also be looked at by considering the meaning of **"if-then"** statements which we first introduced on page twenty one and following. If we assert that p is a sufficient condition for q, then (p $\rightarrow$ q). If p is a necessary condition for q, then q entails p. Thus, if we assert that p is both a sufficient and a necessary condition for q, then we are affirming [(p $\rightarrow$ q) & (q $\rightarrow$ p)] or that p is a bi-conditional of q (p $\leftrightarrow$ q). (p if q), is symbolized as (q $\rightarrow$ p). p only if q, is symbolized as (p $\rightarrow$ q). Thus, if we affirm p if and only if q, we are affirming that p is a bi-conditional of q (p ↔ q) which is logically equivalent to [(p → q) & (q → p)] and also logically equivalent to [(p & q) V (~ p & ~ q)].

Let us try illustrating this by a concrete situation. Let us suppose that we are close to final exam time and you are attempting to figure out what it is going to take for you to get a B in history (H), calculus (C) and English (E) respectively. Let us further suppose that your average in history is such that if you get either an A or B on the final that you will get a B for this course. In this case, getting an A is a sufficient condition for a B in history. (Af → Bh). Getting an A is not a necessary condition, however. You could get either an A or a B and still get your B in history. In English, however, you have a high average but not as high as in history and you also have a paper you have not turned in. Your grade here is such that you must both get an A on the final and pass in the paper. Here an A on the final is a necessary condition for getting a B in the course but not a sufficient condition. You also have to pass in that paper. Hence, (Be → Af). The situation in calculus is different. You must get an A on the final to get a B for the course. But you have turned in all your other homework so that getting an A on the final is the only thing you have to do. Getting an A on the final is enough or is sufficient to get you that B for the course. Thus with respect to calculus, an A on the final is both a necessary and sufficient condition for getting a B for the course. You will get a B for the course, if and only if you get an A for the final. It will be the case either that you get an A on the final and B for the course, or that you both will not get an A on the final and you will not get a B for the course. If you get an A on the final you will get a B for the course and if you get a B for the course you will get an A on the final. You will get an A on the final only if you get a B for the course and you will get a B for the course only if you get an A on the final. In other words, an A on the final is a bi-conditional of a B for the course. (Af ↔ Bc).

Truth Tables For Material Equivalence

Figure 28.1

(p ↔ q) ↔ [(p → q) & (q → p)]

p	q	(p	↔	q)	↔	[(p	→	q)	&	(q	→	p)]
1	1	1	1	1	1	1	1	1	1	1	1	1
1	0	1	0	0	0	1	0	0	0	0	1	1
0	1	0	0	1	1	0	1	1	0	1	0	0
0	0	0	1	0	1	0	1	0	1	0	1	0

Figure 28.2

(p ↔ q) ↔ [(p & q) V (~p & ~q)]

p	q	(p	↔	q)	↔	[(p	&	q)	V	(~p	&	~q)]
1	1	1	1	1	1	1	1	1	1	0	0	0
1	0	1	0	0	1	1	0	0	0	0	0	1
0	1	0	0	1	1	0	0	1	0	1	0	0
0	0	0	1	0	1	0	0	0	1	1	1	1

Flow Charts For Material Equivalence

Figure 28.3

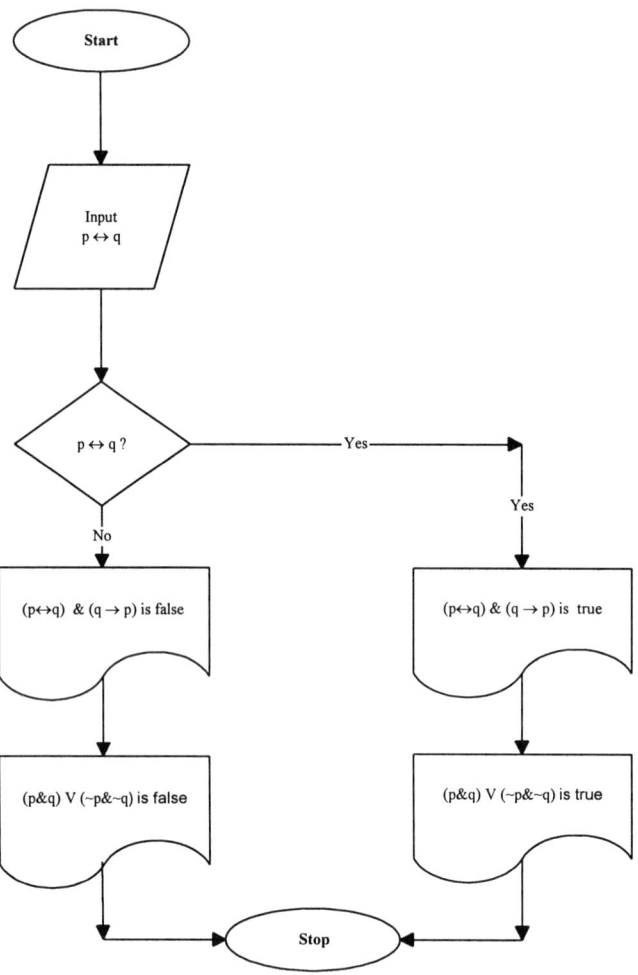

Figure 28.4

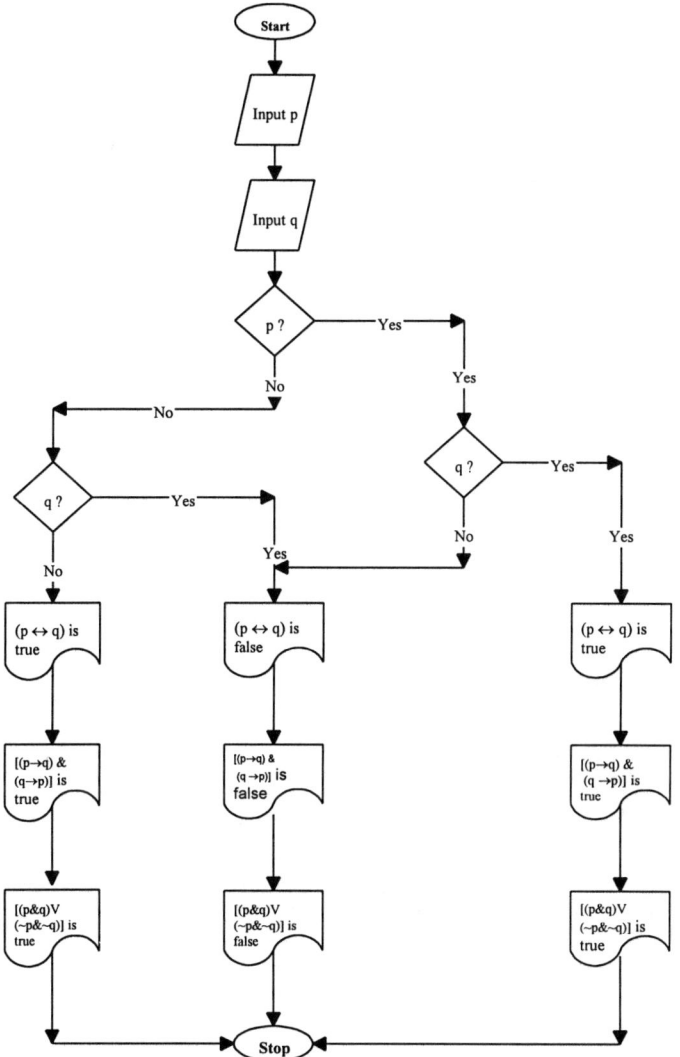

Algorithms For Material Equivalence

Figure 28.5

1. Input value of p ↔ q.
2. If p ↔ q is true,
    (a) [(p →q) & (q → p)] & [(p & q) V (~p & ~q)] is true.
    (b) Stop.
3. [(p →q) & (q → p)] & [(p & q) V (~p & ~q)] is false.
4. Stop.

Figure 28.6

1. Input value of p.
2. Input value of q.
3. If p is true,
    (a) If q is true, (p ↔ q) & [(p → q) & (q → p)] &
         [(p & q) V (~p & ~q)] are true.
    (b) Stop.
4. If q is false,
    (a) (p ↔ q) & [(p → q) & (q → p)] &
         [(p & q) V (~p & ~q)] are true.
    (b) Stop.
5. (p ↔ q) & [(p → q) & (q → p)] &
    [(p & q) V (~p & ~q)] are false.
6. Stop.

It will be helpful to remember when attempting to solve the problems below that there are two definitions for p ↔ q. [(p → q) & (q → p)] and [(p & q) V (~p & ~q)] are both legitimate replacements for (p ↔ q). In solving a particular problem one definition may work where the other will not.

PROBLEM SET: MATERIAL EQUIVALENCE
Demonstrate the validity of the following arguments.

1.
    1.   D ↔ C   ∴   C → D

2.
    1.   D ↔ C   ∴   ~C
    2.   ~(D & C)

3.
    1.   D ↔ C   ∴   ~D
    2.   ~C

4.
    1.   ~C V D   ∴   D ↔ C
    2.   ~C → ~D

5.
    1.   ~D   ∴   D ↔ C
    2.   ~C

    6.
    1.   ~(~C & ~D) → (D & C)   ∴   D ↔ C

7.
    1.   [~A V (B V C)] & [~B V (A V C)]   ∴ A ↔ B
    2.   ~C

8.
    1.   A → C   ∴   A ↔ C
    2.   ~F V (~A & ~C)
    3.   A V F

9.
    1.   C V D   ∴   [(A ↔ B) & (F ↔ G)] V (D ↔ D)
    2.   S V ~R
    3.   (R → S) → (A & B)
    4.   ~(~G & ~F) → ~D
    5.   (F & G) V ~C

10.
    1.   A ↔ B   ∴   ~B V (A & B)

1. Skinner is right if and only if behaviorism is correct. Therefore, if behaviorism is correct then Skinner is right.   S, B.

2. Freud is right only if sexual instinct is one of the primary determinants of human behavior. Sexual instinct being one of the primary determinants of human behavior is a sufficient condition for Freud being right. Therefore, Freud being right is a bi-conditional of sexual instinct being one of the primary determinants of human behavior.   F, S

3. If Freud is right then, sexual instinct is one of the primary determinants of human behavior and the instinct of aggression is one of the primary determinants of human behavior. If the instinct of aggression is one of the primary determinants of human behavior then, sexual instinct is one of the primary determinants of human behavior only if Freud is right. Therefore, Freud's being right is a bi-conditional of sexual instinct and the instinct of aggression both being primary determinants of human behavior.   F, S, A.

4. God is omnipotent. God is also loving. Therefore, God is omnipotent if and only if God is loving.   O, L.

5. Man is either free or morally perfect. If he is morally perfect then he is not free. Therefore, man's not being free is materially equivalent to his being perfect.   F, P.

6. David was a priest if and only if he was a Levite. David was not both a priest and a Levite. Therefore, David was not a Levite.   P, L.

7. Either David was not a priest or he was a Levite. David was not a priest only if he was not a Levite. Therefore, David was a priest if and only if he was a Levite.   P, L.

8. David was not the son of Saul. David was not the brother of Jonathan, Therefore, David was the son of Saul if and only if David was the brother of Jonathan.   S, J.

9. It is not the case that David's not the brother of Jonathan and not being the son of Saul, is a sufficient condition for David's being the son of Saul and the brother of Jonathan. Therefore, David being a son of Saul is a biconditional of David being a brother of Jonathan. J, S.

10. If Augustine is right, then God transcends time. Either Whitehead is not right, or Augustine is not right and God does not transcend time. Augustine is right or Whitehead is right. Therefore, Augustine is right if and only if God transcends time. A, T, W.

CHAPTER 29:   DISTRIBUTION
[p & (q V r)]↔[(p & q) V (p & r)]
[p V (q & r)]↔[(p V q) & (p V r)]

As with the other rules of replacement it will help to fix the principle of distribution (abb. Dist.) in the mind if you mentally "think through" the principle rather than simply trying to memorize it.

The first version asserts that if p is true and if q V r is also true, then either (p & q) V (p & r) must be true. Why must this be so? The explanation is as follows. If I have a p, then if q V r is true, then I have a q or I have an r. If I have a q, then p & q is true. If I have an r, then p & r is true. Thus, I have either (p & q) V (p & r).

Lets work the same relationship form right to left. In order to have p & q, I must have a p. In order to have p & r, I must have a p. Therefore, if I have (p & q) V (p & r), I must have a p. If I have a p, then we know (by principle of addition) that p or anything else including p V(q & r) is true.

We can "think through" the second version of distribution in a similar way. The left hand statement asserts [p V (q & r)]. It is clear that if the p statement is true then both (p V q) and (p V r) would be true. If on the other hand, q & r is true, then since q would be true (p V q) would be true and since r would be true (p V r) would be true. Thinking the relationship through from right to left yields similar results. The right hand statement of the relationship asserts [(p V q) & (p V r)]. If p is true then [p V (q & r)] would have to be true. On the other hand, if p is false then for (p V q) to be true, q must be true and for (p V r) to be true, r must be true. This means that (q & r) is true. If (q & r) is true, then of course [p V (q & r)] has to be true. Thus, the values of p, q and r can be distributed in the various different ways specified without changing their logical relationships to each other.

Flow Charts For Distribution

$$[p \& (q \lor r)] \leftrightarrow [(p \& q) \lor (p \& r)]$$

Figure 29.1

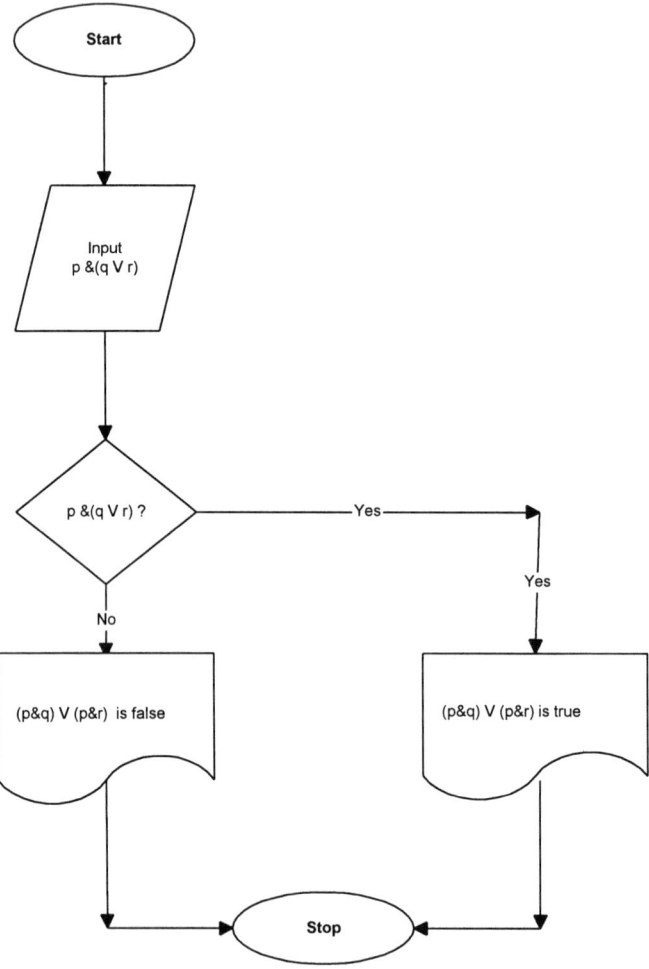

Figure 29.2

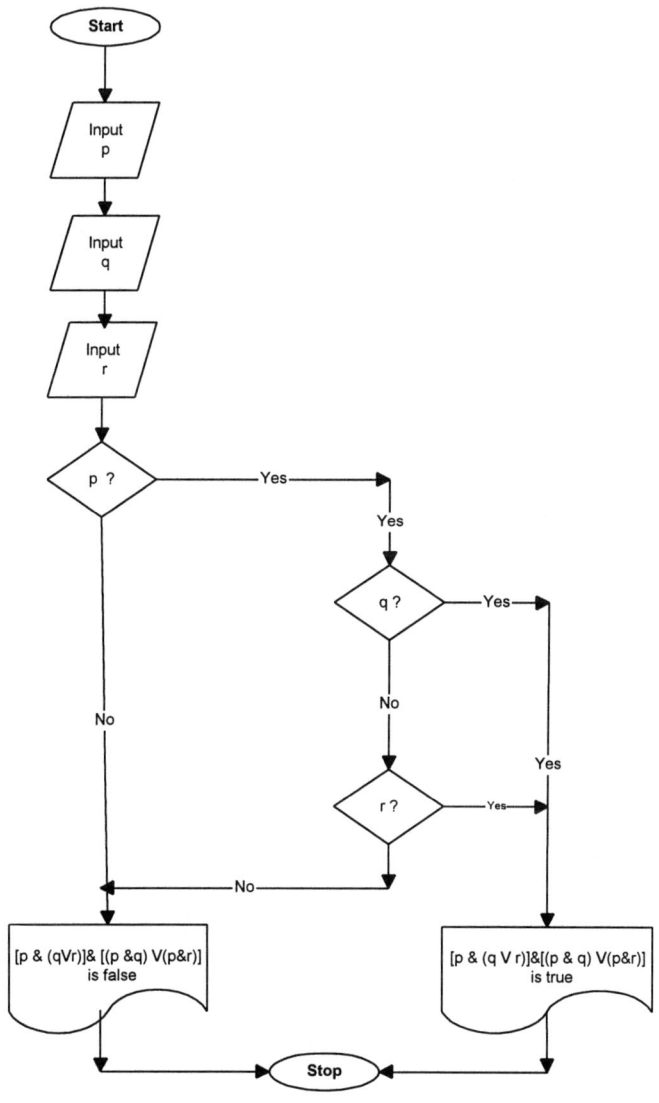

$$[p \lor (q \& r)] \leftrightarrow [(p \lor q) \& (p \lor r)]$$

Figure 29.3

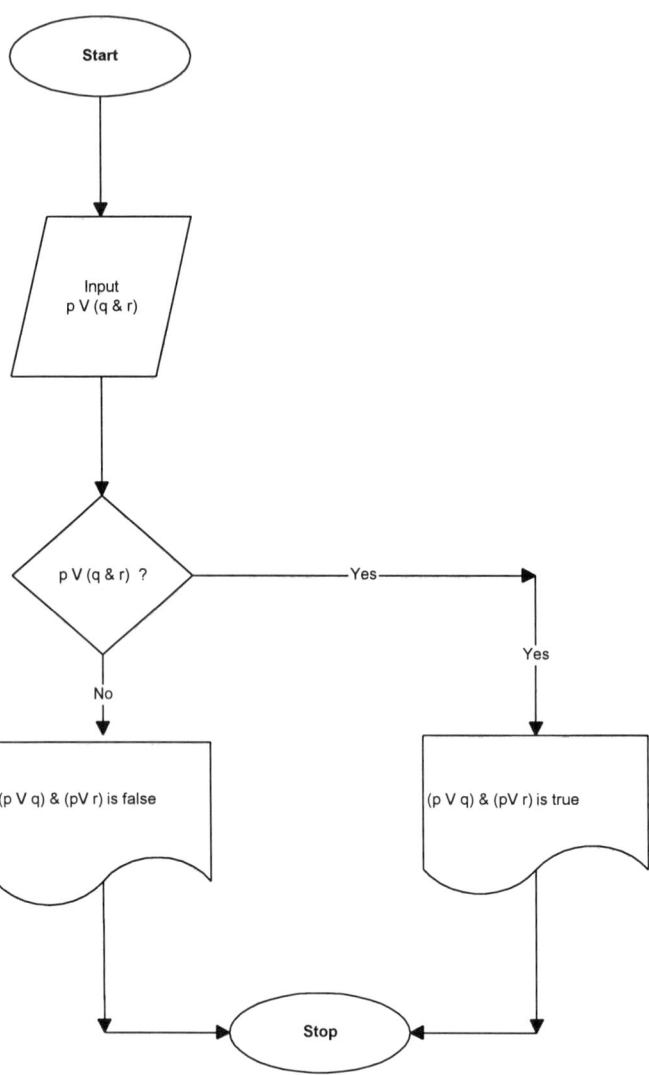

Figure 29.4

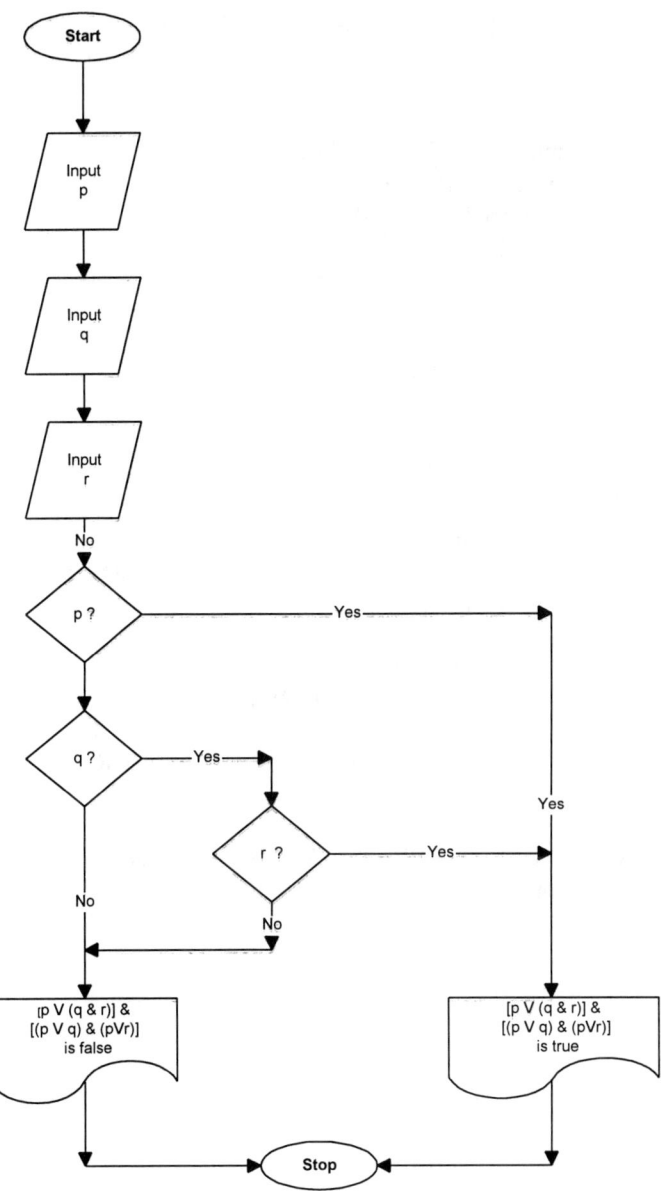

Algorithms For Distribution

$$[p \& (q \lor r)] \leftrightarrow [(p \& q) \lor (p \& r)]$$

Figure 29. 5

1. Input value of p & (q V r).
2. If p & (q V r) is true,
    (a) (p & q) V (p & r) is true.
    (b) Stop.
3. (p & q) V (p & r) is false.
4. Stop.

Figure 29.6

1. Input value of p.
2. Input value of q.
3. Input value of r.
4. If p is false,
    (a) [p & (q V r)] & [(p & q) V (p & r)] are false.
    (b) Stop.
5. If q is false,
    Then if r is false,
    (a) [p & (q V r)] & [(p & q) V (p & r)]
        are false.
    (b) Else [p & (q V r)] and [(p & q) V (p & r)]
        are true.
    (c) Stop.
6. [p & (q V r)] & [(p & q) V (p & r)] are true.
7. Stop.

$$[p \lor (q \ \& \ r)] \leftrightarrow [(p \lor q) \ \& \ (p \lor r)]$$

Figure 29.7

1.   Input value of p ∨ (q & r).
2.   If p ∨ (q & r) is true,
      (a) [(p ∨ q) & (p ∨ r)] is true.
      (b) Stop.
3.   [p ∨ q) & (p ∨ r)] is false.
4.   Stop.

Figure 29.8

1.   Input value of p.
2.   Input value of q.
3.   Input value of r.
4.   If p is true,
      (a)[p ∨ (q & r)] and [(p ∨ q) & (p ∨ r)] are true.
      (b) Stop.
5.   If q is true,
      Then if r is true,
      (a) [p ∨ (q & r)] & [(p ∨ q) & (p ∨ r)] are true.
      Else (b) [p ∨ (q & r)] & [(p ∨ q) & (p ∨ r)] are false.
      (c) Stop.
6.   [p ∨ (q & r)] & [(p ∨ q) & (p ∨ r)] are false.
7.   Stop.

PROBLEM SET: DISTRIBUTION
Demonstrate the validity of the following arguments.

1.
   1.   D & (B ∨ A)   ∴ (A & D) ∨ (B & D)

2.
   1.   D ∨ (B & A)   ∴ D
   2.   ~ A

3.
  1. E V (F & G)   ∴ A ↔ B
  2. ~ F V (A & B)
  3. (~ B & ~ A) V ~ E

4.
  1. E V (F & G)   ∴ A ↔ B
  2. ~ F V (A & B)
  3. (~ B & ~ A) V ~ E

5.
  1. A V [(F & G) & H]   ∴ A
  2. ~ F

6.
  1. [(~ A & B) & (E & F)] V (I & H) ∴ A → I

7.
  1. A → B   ∴ (~ A V A) & (~ A V B)

8.
  1. D → (C & F)   ∴ [(D →C) & (D → F)] & (A↔ B)
  2. B V ~ A
  3. B→ A

9.
  1. B V (~ A & ~ D) ∴ (A ↔ B) & (B ↔ D)
  2. ~ B V (A & D)

10.
  1. ~ H   ∴ G V (J & I)
  2. G V (H & I)

11.
  1. ~ A & D   ∴ C
  2. (D & C) V (A & B)

1. Hezbollah is a terrorist organization, and its members are either Shiite Muslims or they are Sunnite Muslims. Therefore, Hezbollah is a terrorist organization and its members are Shiite Muslims, or Hezbollah is a terrorist organization and its members are Sunnite Muslims.

    Let T = Hezbollah is a terrorist organization.
    Let S = its members are Shiite Muslims.
    Let M = its members are Sunnite Muslims.

2. Hezbollah is a terrorist organization, or it is a political party and primarily a religious worshipping community. Hezbollah is not primarily a religious worshipping community. Therefore, Hezbollah is a terrorist organization. T, P, R.

3. If Henry II was the husband of Eleanor of Acquitaine, then he was the father of Richard the Lion Hearted and he was the father of the King John who signed the Magna Carta at Runnymede in 1215 A.D. If Henry was the father of Richard the Lion Hearted, then he was the husband of Eleanor of Acquitaine and the one responsible for the death of Becket. Therefore, Henry II was the husband of Eleanor of Acquitaine, if and only if, he was the father of Richard the Lion Hearted.   E, R, J, B

4. Henry II was ruler of Normandy only if he was ruler of England. Therefore, Henry II was either not ruler of Normandy or he was ruler of Normandy, and Henry II either was not ruler of Normandy or he was ruler of England. N, E.

5. Henry II was King of England and ruler of half of France; or he was ruler of Normandy and he was ruler of Brittany, or he was not ruler of Acquitaine. It is not the case that he was not ruler of Acquitaine. Therefore, Henry II was not ruler of England only if he was ruler of Normandy. E, F, N, B, A

6. If Durkheim is right, then capitalism leads to an increase in anomic suicide. Either Durkheim is not right, or capitalism leads to an increase in egoistic suicide. Therefore, Durkheim is not right, or capitalism leads to an increase in anomic suicide and an increase in egoistic suicide.
    D, A, E.
    Note: Anomic suicide is associated with a decline in social support due to a breakdown in social norms. Egoistic suicide is associated with a decline in social support due to an excessive emphasis on individualism.

7. If Marx is right, then social solidarity will not increase among the bourgeoisie and will increase among the proletariat. Social solidarity will increase among the bourgeoisie. Therefore, Marx is not right.
    M, B, P.

8. If Augustine is right, then history can appropriately be understood as God's sovereign working and history can appropriately be understood as a conflict between two kingdoms. If social Darwinism is right, then history can be appropriately understood as part of a natural evolutionary process and history can be appropriately understood as representing a survival of the fittest. Augustine not being right is a bi-conditional of Social Darwinism being right. History cannot appropriately be understood as a survival of the fittest. Therefore, history can appropriately be understood as a conflict between two kingdoms.

     Let A = Augustine is right.

     Let G = history can appropriately be understood as God's sovereign working.

     Let K = history can appropriately be understood as a conflict between two kingdoms.

     Let S = Social Darwinism is right.

     Let E = history can appropriately be understood as part of a natural evolutionary process.

     Let F = history can appropriately be understood as a survival of the fittest.

9. Jim is guilty and Sam is his accomplice, and Mary is guilty and Louise is her accomplice; or Ray is an undercover cop and Adam is his partner. Therefore, if Ray is not an undercover cop, then Louise is Mary's accomplice. J, S, M, L, R, A

10. It is not the case that, Jim is guilty and Sam is his accomplice, if and only if, Mary is guilty and Louise is her accomplice. Therefore, Mary is not guilty only if Jim is guilty. J, S, M, L.

CHAPTER 30: De MORGAN'S THEOREMS

$$\sim(p \ \& \ q) \leftrightarrow (\sim p \ V \sim q)$$
$$\sim(p \ V \ q) \leftrightarrow (\sim p \ \& \sim q)$$

De Morgan's Theorems (abb. DeM.) are named after their discover, the British mathematician and logician Augustus De Morgan (1806-1871).   As with the other rules of replacement, the best procedure is to "think through" the relationships before attempting their memorization. Both of these theorems become clear with a little reflection. The first theorem asserts $\sim(p \ \& \ q) \leftrightarrow (\sim p \ V \sim q)$. The left hand side of the relationship denies that the conjunction of p & q is true. Based on our understanding of the meaning of the conjunction symbol (&) and the inclusive or symbol (V) we can readily see the truth of the bi-conditional assertion being made (p & q) is true only if both p and q are true. If either is false, then it is not the case that they are both true which we symbolize as $\sim(p \ \& \ q)$. This is what the right hand side of the bi-conditional asserts. $(\sim p \ V \sim q)$ asserts that either (1) p is false, or (2) q is false, or (3) both p and q are false. In any case, both of them are not true.

The second theorem: $\sim(p \ V \ q) \leftrightarrow (\sim p \ \& \sim q)$ makes a different claim but can be treated in a somewhat similar fashion. The left hand side of the theorem asserts that it is not the case that either p or q is true. Now if neither is true, then it must be the case that both are false which is what the right hand side of the theorem asserts. This works both ways.

If we know that they both are false we know that it is not the case that either one is true.

The Application of Rules for De Morgan's Theorems

How may we apply De Morgan's Theorems when the exact form for doing so is not already in place? There are at least a couple of ways of going about this. One way uses the principle of double negation (D.N.) to create that proper form. Another less cumbersome way dispenses with D. N. In this text we will use the second method although this is not the standard procedure in most texts.

Rules for De Morgan Transformations

1. If there are one or more negation signs in front of the parenthesis of the combined variables subtract one. Otherwise add a negation sign. In other words, either subtract or add a negation sign.

2. If there are one or more negation signs in front of any of the variables within the parenthesis subtract one. Otherwise add a negation sign. In other words, either subtract or add a negation sign in front of each variable.

3. If you are starting with a conjunction sign (&) change it to an inclusive or sign (V). If you are starting with an inclusive or sign (V), change it to a conjunction sign. In other words, either change & to V, or change V to &.

Figure 30.1

Examples of application of three rules.

#1	1. p & q 2. ~(~p V ~q) 1, DeM.	#3	1.~p & ~~q 2.~(p V ~q) 1, DeM.
#2	1. p V q 2. ~(~p & ~q) 1, DeM.	#4	1.~~(~~p V ~q) 2.~(~p & q) 1, DeM.

Truth Tables For De Morgan's Theorems

Figure 30.2

~(p & q) ↔ (~p V ~q)

p	q	~	(p	&	q)	↔	(~	p	V	~	q)
1	1	**0**	1	1	1	1	0		**0**	0	
1	0	**1**	1	0	0	1	0		**1**	1	
0	1	**1**	0	0	1	1	1		**1**	0	
0	0	**1**	0	0	0	1	1		**1**	1	

Figure 30.3

~(p V q) ↔ (~p & ~q)

p	q	(~	p	V	~	q)	↔	~	(p	&	q)
1	1	0		**0**	0		1	**0**	1	1	1
1	0	0		**1**	1		1	**1**	1	0	0
0	1	1		**1**	1		1	**1**	0	0	1
0	0	1		**1**	1		1	**1**	0	0	0

Flow Charts For De Morgan's Theorems

$$\sim(p \And q) \leftrightarrow (\sim p \lor \sim q)$$

Figure 30.4

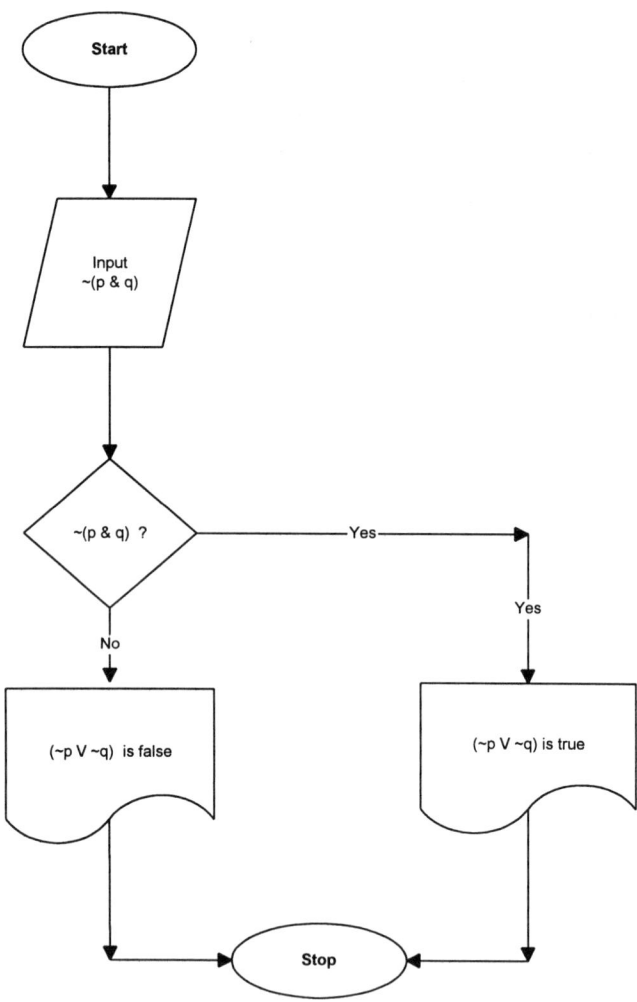

Figure 30.5

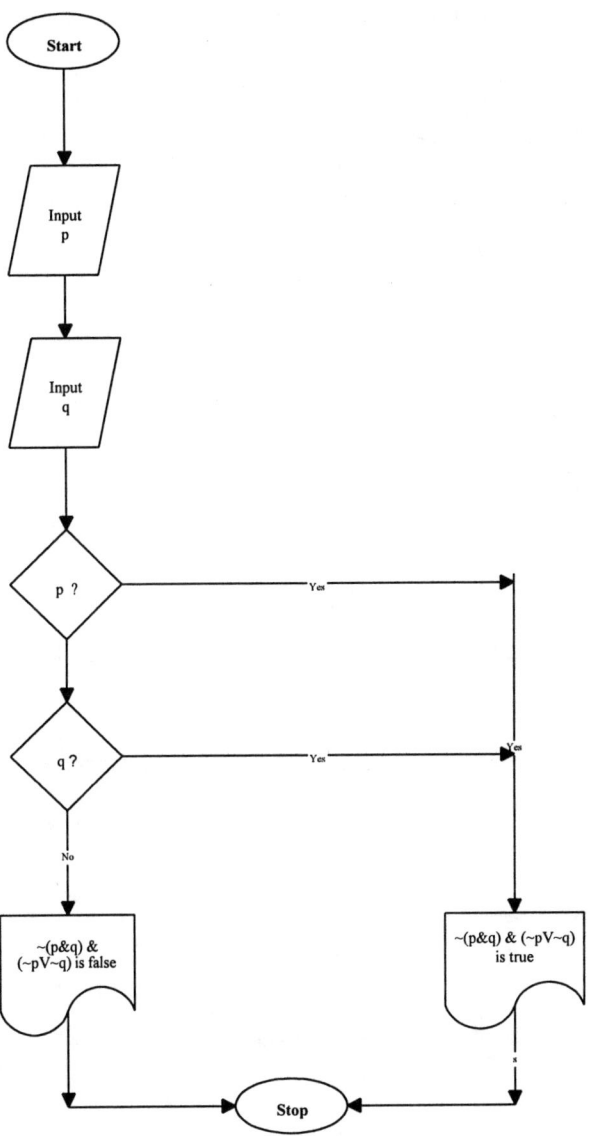

Logic of the Computing Sciences

$$\sim(p \lor q) \leftrightarrow (\sim p \;\&\; \sim q)$$

Figure 30.6

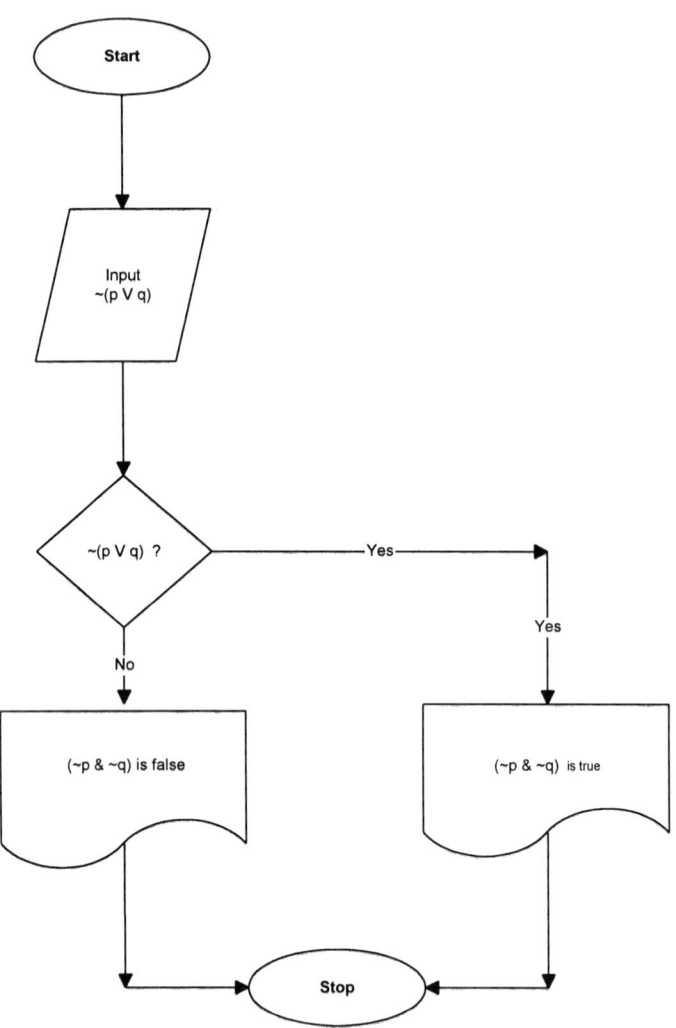

Figure 30.7

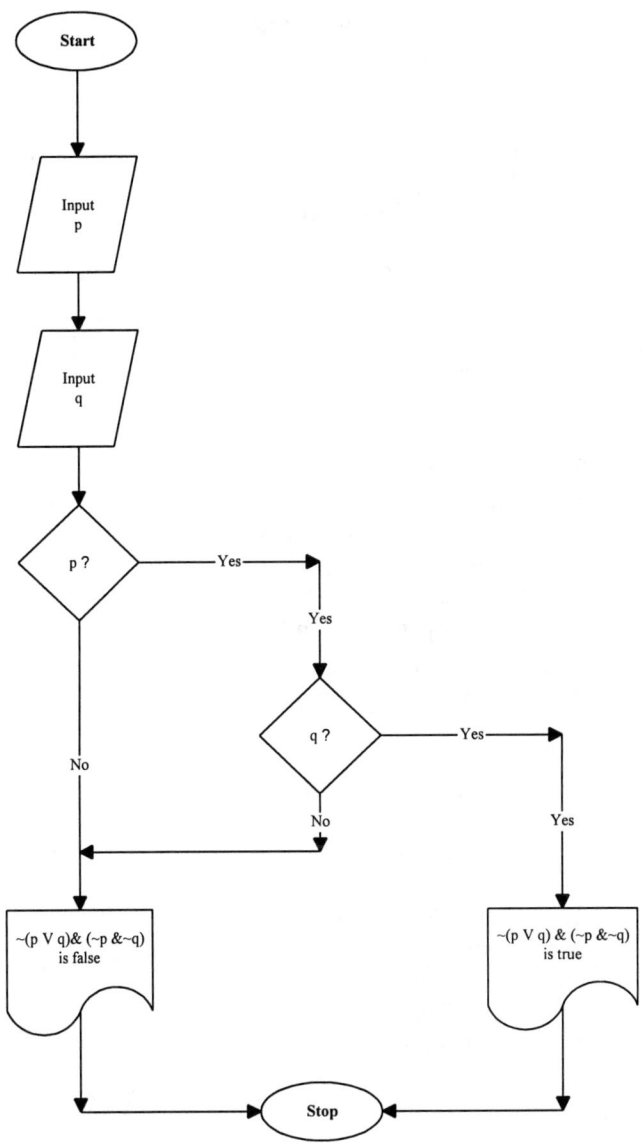

Algorithms For De Morgan's Theorem

$$\sim(p \ \& \ q) \leftrightarrow (\sim p \ V \sim q)$$

Figure 30.8

1.   Input value of $\sim(p \ \& \ q)$.
2.   If $\sim(p \ \& \ q)$ is true,
         (a) $(\sim p \ V \sim q)$ is true.
         (b) Stop.
3. $(\sim p \ V \sim q)$ is false.
4.   Stop.

Figure 30.9

1.   Input value of p.
2.   Input value of q.
3.   If p is false,
         (a) $\sim(p \ \& \ q)$ and $(\sim p \ V \sim q)$ are true.
         (b) Stop.
4.   If q is false,
         (a) $\sim(p \ \& \ q) \ \& \ (\sim p \ V \sim q)$ are true.
         (b) Stop.
5. $\sim(p \ \& \ q) \ \& \ (\sim p \ V \sim q)$ are false.
6.   Stop.

$$\sim(p \ V \ q) \leftrightarrow (\sim p \ \& \sim q)$$

Figure 30.10

1.   Input value of $\sim(p \ V \ q)$.
2.   If $\sim(p \ V \ q)$ is true,
         (a) $(\sim p \ \& \sim q)$ is true.
         (b) Stop.
3. $(\sim p \ \& \sim q)$ is false.
4.   Stop.

Figure 30.11

1. Input value of p.
2. Input value of q.
3. If p is true,
    (a) ~(p V q) and (~p & ~q) are false.
    (b) Stop.
4. If q is true,
    (a) ~(p V q) & (~p & ~q) are false.
    (b) Stop.
5. ~(p V q) and (~p & ~q) are true.
6. Stop.

PROBLEM SET: DE MORGAN'S THEOREM
Demonstrate the validity of the following arguments.

1. 1. $(X \& Y) \rightarrow D$ $\therefore \sim Y \vee \sim X$
   2. $\sim D$

2. 1. $\sim A$ $\therefore \sim (A \vee B)$
   2. $B \rightarrow A$

3. 1. $P \rightarrow Q$ $\therefore \sim (P \& \sim Q)$

4. 1. $\sim (P \rightarrow Q)$ $\therefore P \& \sim Q$

Problems number three and four give us an interesting, if predictable (upon reflection) commentary on the meaning of $(p \rightarrow q)$ and $\sim (p \rightarrow q)$ respectively. Through Imp.; D.N. and DeM. we are able to derive $\sim (p \& \sim q)$ from $\sim (p \rightarrow q)$ in problem number three, and $p \& \sim q$ from $\sim (p \rightarrow q)$ in problem number four. If we examine the truth table for $p \rightarrow q$ we see that $p \rightarrow q$ is false only in logical possibility number two, that situation where p is true and q is false. The assertion that $p \rightarrow q$ is true, therefore is tantamount to the denial of logical possibility number two or $\sim (p \& \sim q)$. The reverse is the case for $\sim (p \rightarrow q)$ (IMP; DeM.; D.N). Denying that $p \rightarrow q$ is true involves the affirmation of that condition under which $p \rightarrow q$ is false, namely logical possibility number two in our truth table. Thus, $\sim (p \rightarrow q)$ entails the affirmation of $p \& \sim q$.

5.
1. $F \rightarrow \sim B$ $\therefore \sim (A \mathbin{\&} B)$
2. $\sim E \rightarrow F$
3. $E \rightarrow \sim A$

6.
1. $\sim [\sim (A \mathbin{\&} B) \mathbin{\&} \sim (A \mathbin{\&} C)]$ $\therefore B$
2. $\sim C$

7.
1. A $\therefore \sim[\sim (A \vee B) \vee \sim (A \vee C)]$

8.
1. $\sim A$ $\therefore \sim [(A \vee B) \mathbin{\&} (C \vee D)]$
2. $\sim B$

9.
1. $\sim\{[(\sim A \vee \sim B) \vee \sim C] \vee [(\sim D \vee \sim E) \vee \sim F]\}$
                              $\therefore [F \mathbin{\&} (E \mathbin{\&} D)] \mathbin{\&} [C \mathbin{\&} (B \mathbin{\&} A)]$

10.
1. $A \rightarrow C$ $\therefore \sim[(A \vee B) \mathbin{\&} \sim C]$
2. $B \rightarrow C$

1. Marx and Spencer are not both right. Therefore, either Marx is not right or Spencer is not right. M, S.

2. Neither Marx or Spencer is right. Therefore, Marx is not right and Spencer is not right. M, S.

3. It is not the case both that Marx is not right and that Spencer is not right. Therefore, if Marx is not right, then Spencer is right.   M, S.

4. If Carl's testimony is true and Richard's testimony is true, then Sam will go to jail and Tom will go to jail. Sam will not go to jail. Therefore, Carl's testimony being true entails that Richard's testimony is not true.
                           C, R, S, T.

5. It is not true both that John loves Susan and that he will not ask Susan to marry him. It is also not true both that John will ask Susan to marry him and that he does not love her. Therefore, John loves Susan, if and only if he will ask her to marry him.
     Let L = John loves Susan.
     Let M = John will ask Susan to marry him.

6.   From Leviticus Chapter 11:
     If an animal does not both chew its cud and have a split hoof, then it is not either ceremonially clean or permissible to eat. Therefore, if the animal is permissible to eat, then it chews its cud.

       Let C = the animal chews its cud.
       Let S = the animal has a split hoof
       Let A = the animal is ceremonially clean.
       Let B = the animal is permissible to eat.

7.   If the proletariat is both exploited and becomes aware of its situation, then it will not occur that the classless society will not be established or that capitalism will not be destroyed. Therefore, if capitalism is not destroyed then it will not occur both that the proletariat is exploited and that the proletariat becomes aware of its situation.

       Let P = the proletariat is exploited.
       Let A = the proletariat will becomes aware of its situation.
       Let C = the classless society will be established.
       Let D = capitalism will be destroyed.

8.   The following is not true: either interest rates will rise and the stock market will not fall, or the stock market will fall and interest rates will not rise. Therefore, interest rates will rise, if and only if, the stock market falls. I, S.

9.   The following is not the case: Its not true that either Jim is guilty or Sam is not guilty; or its not true both that Sam or Harry is guilty, and Jim is not guilty. Therefore, Jim is guilty, if and only if, Sam is guilty.
                                     J, S, H.

10.   Either it is not true both that Jim is guilty and Harry is guilty, or it is not true that either Jim or Harry is guilty. Therefore, if Jim is guilty, then Harry is not guilty. J, H.

CHAPTER 31: QUANTIFICATION THEORY

Up until now we have been considering logical problems in which the validity or soundness of the argument depends upon how simple statements are compounded. For instance, if the simple statement "Suzy is intelligent" is true and the simple statement "John is intelligent" is true, then the compound statement "Suzy is intelligent and John is intelligent" can be shown to be true by the principle of conjunction. For another example, we might consider the compound statement "Lincoln was the first president or Lincoln was assassinated." This compound statement is true because one of the simple statements comprising it, namely "Lincoln was assassinated" is true and this is all that is required for an "inclusive or" compound statement to be true.

But let us consider the following obviously valid argument which contains no compound statement at all.

All Lutherans are Christians. Jones is a Lutheran. Therefore, Jones is a Christian.

We might symbolize this as is shown in figure 31.1

Figure 31.1

$$
\begin{array}{c}
L \\
J \\
\hline
\therefore \quad C
\end{array}
$$

Though obviously valid, the statement cannot be demonstrated to be valid by the techniques we have developed thus far. In order to bring the argument into a form closer to the techniques we have been using we may translate the statement as follows: "All Lutherans are Christians" appears to equal "If anything is a Lutheran, then it is a Christian." We might symbolize it as Lc. But this doesn't quite capture all that we want to say. We are saying that of all things in the universe, if you can call it Lutheran, then you can call it Christian.

Or to put it another way, if being a Lutheran can be predicated of something, then being Christian can be predicated of that thing. Let x stand for anything in the universe. Then we may say, for any x if x is a Lutheran, then x is a Christian. We may symbolize this as follows: (x)(Lx → Cx) For any X if L of x, then C of x.

If we wanted to symbolize the statement "No Lutherans are Christians" we would do so as follows: (x)(Lx → ~Cx) or "for anything in the universe, if being a Lutheran can be predicated of that thing, then **not** being Christian may be predicated of that thing."

Now both of these statements **could** be true even if there were no Lutherans and no Christians in the universe. Just as all unicorns are one horned animals is true (x)(Ux → Ox) when we see that it means that if a thing is a unicorn, then it must be a one horned animal. This can be shown by the truth table for if p then q (p → q).

Figure 31.2

	p	→	q
1.	1	1	1
2.	1	0	0
3.	0	1	1
4.	0	1	0

In logical possibility #3, p does not exist but p → q is true. In logical possibility #4, neither p nor q is assumed to exist but p→ q is still true. This is because these hypothetical statements assert a necessary relationship between two variables but do not assert the **existence** of anything. The propositions do not have what is referred to as **existential import**. The two following statements do have existential import. "Some Lutherans are Christians" and "Some

Lutherans are not Christians." Another way of saying this is that, "there exists in the universe something (an x) such that being Lutheran may be predicated of that x and being Christian may be predicated of that x, or (∃x)(Lx & Cx)." "Some Lutherans are not Christians" may be written, "there exists an x such that being Lutheran may be predicated of that x and not being Christian may be predicated of that x, (∃x)(Lx & ~Cx). The backwards e (∃) is the symbol designating existential import.

Translate the following sentences into symbolic form.*

1.   All Lutherans are Christians.
2.   No Lutherans are Christians.
3.   Some Lutherans are Christians.
4.   Some Lutherans are not Christians.
5.   Only Christians are Lutherans.
6.   Only Lutherans are Christians.
         Let Lx = x is Lutheran;
         Let Cx = x is Christian.)
7.   Some Muslims are Shiites.
8.   Many Muslims are not Shiites.
9.   Only Muslims are Shiites.
10.   No Shiites are Muslims.
         Let Mx = x is Muslim;
         Let Sx = x is Shiite.)

We are now, perhaps in a position to introduce several new principles which will help us to demonstrate the validity of our original argument. This is shown in figure 31.3

---

* It needs to be noted that our symbolic system distinguishes only between all (X) and some (∃X). "Many," "most," "some," "at least one," "almost all," etc. , all equal "some" or (∃X).

Figure 31.3

All Lutherans are Christians.
Jones is a Lutheran.
Therefore, Jones is a Christian.

The above argument may be symbolized as follows.:

1.	(x)(Lx → Cx)	∴ Cj
2.	Lj	
3.	Lj → Cj	1, UI. (Universal Instantiation)
4.	Cj	3,2 M.P.

Steps three and four are made possible by the introduction of the principle known as universal instantiation. (U.I.) The rationale for this is as follows. If it is true that for anything in the universe (any x) that if being Lutheran can be predicated of that x, then being Christian can be predicated of that x; then this relationship must obtain for all particular instances. "If being Lutheran can be predicated of Jones, then being Christian can be predicated of Jones" is a particular instantiation of the general or universal proposition which states that its true of everything that if Lutheran then Christian. What is true for everything in the universe must be true of every particular instance (including Jones).

Let us proceed to a similar but different problem as depicted below.

Figure 31.4

All Lutherans are Christians.
All Christians are Trinitarians.
Therefore, all Lutherans are Trinitarians.

1.	(x)(Lx → Cx)	∴ (X)(Lx → Tx)
2.	(x)(Cx → Tx)	
3.	La → Ca	1, UI
4.	Ca → Ta	2, UI
5.	La → Ta	3, 4, H.S.
6.	(x)(Lx→ Tx)	5, U.G. (Universal Generalization)

The preceding problem introduces the principle of Universal generalization (U.G). Let us analyze the argument step by step. In this problem, both our premises and the conclusion are in the form of formerly non-compound statements. In step #3 we derived La → Ca by 1, UI, where **a** stands for a **representative** and arbitrarily chosen individual and the entire statement is formerly a compound statement (to which our nineteen rules of inference may apply). The key term here is **representative**. Because (a) is treated as a particular instance which is **representative** of all instances, we may substitute it for (x) (the letter designating all). Likewise because, (a) is **representative** of all instances once we have derived La → Ta by 3, 4, H.S. we may in step five generalize to (x)(Lx → Tx) by 5, U.G.

Let us look at a different problem as depicted in figure 31.5

Figure 31.5

All Lutherans are Christians.
Some Lutherans are fat.
Therefore, some Christians are fat.

1.	(x)(Lx → Cx)	∴ (∃ x)(Cx & Fx)
2.	(∃ x)(Lx & Fx)	
3.	La & Fa	2, EI (Existential Instantiation)
4.	La → Ca	1, UI
5.	La	3 Simp.
6.	Ca	4, 5 M.P.
7.	Fa & La	3, Com.
8.	Fa	7, Simp.
9.	Ca & Fa	6,8, Conj.
10.	(∃x)(Cx & Fx)	9 EG (Existential Generalization)

The immediately preceding problem introduces two new principles Existential Instantiation (E.I.) and Existential Generalization (E.G.) In step #3 we infer from premise #2 that La & Fa are true. Here (a) is an arbitrarily chosen individual designation which **has not been used** before in the problem. In step #4, we use UI and then proceed from #5 through #8 using familiar rules of inference. In #10 we introduce the rule of Existential Generalization. Here we reason that since C may be predicated of a, and F may be predicated of a, that there exists in the universe at least one thing such that C may be predicated of that thing and F may be predicated of that thing ($\exists$x)(Cx & Fx).

When applying the rule of Existential Instantiation it is important that we use the rule designating individuals that have not been previously used in the problem. For example, if we ignore this rule we could deduce the following contradictory results from the premises as shown in figure 31.6.

Figure 31.6

1.	($\exists$ x)(Lx & Fx)	$\therefore$ (Lj & ~Lj)
2.	($\exists$ x)(~Lx & ~Fx)	
3.	Lj & Fj	1 EI
4.	~ Lj & ~ Fj	2 EI (illegitimate)
5.	Lj	3 Simp.
6.	~ Lj	4 Simp.
7.	Lj & ~ Lj	5,6 Conj

This might be translated as being Lutheran may be predicated of Jones and it is not the case that being Lutheran may be predicated of Jones. To avoid this error we need to use another letter in step #4 since we had already used j in Existential Instantiation in step #3.

PROBLEM SET: QUANTIFICATION THEORY

1.
    1. (x)(Dx $\rightarrow$ Fx) $\therefore$ (x)(Dx $\rightarrow$ Cx)
    2. (x)(Fx $\rightarrow$ Cx)

2.
    1. (x)(Dx $\rightarrow$ Fx) $\therefore$ Fa
    2. Da

3.
    1. (x)(Dx$\rightarrow$ Fx) $\therefore$ ($\exists$x)(Fx & Rx)
    2. ($\exists$x)(Dx & Rx)

4.
    1. (x)(Dx $\rightarrow$ Fx) $\therefore$ ($\exists$x)(~Dx & Px)
    2. ($\exists$x)(~Fx & Px)

5.
1. (x)(Dx → Fx)   ∴ (∃x)(Px & Rx)
2. (x)(~Dx → Rx)
3. (∃x)(~Fx & Px)

6.
1. (x)(~Gx → ~Fx)   ∴ (x)(Fx → Hx)
2. (x)(~Hx → ~Gx)

7.
1. (x)(Ax → Bx)   ∴ (x)(~Ex → Ax)
2. (x)(~Cx → ~Bx)
3. (x)(Cx → Dx)
4. (x)(~Ex → ~Dx)

8.
1. Ie V Ke   ∴ ~Me → Ne
2. (x)(~Mx → ~Ix)
3. (x)(~Nx → ~Kx)

9.
1. (x)(Bx → Cx)   ∴ ~(~Ce V ~Ee)
2. (x)(Dx → Ex)
3. Be & De

10.
1. Ae & Be   ∴ ~(Ae → ~Ce)
2. (x)(Bx → Cx)

1. Only Muslims are Shiites. John is a Shiite. Therefore, John is a Muslim.
    M, S, j.

2. John is a Muslim. He is also a Shiite. Therefore, there exists at least one
    individual who is both a Muslim and a Shiite. M, S, j.

3. All Shiites are Muslims. Therefore, John is a Shiite only if he is a
    Muslim. S, M, j.

4. All attorneys are honest men. All honest men are poor. Therefore, all
    attorneys are poor. A, H, P.

5. All social Darwinists are politically conservative. Therefore, all non
    political conservatives are not social Darwinists. S, P

6. All who have not passed the bar are not attorneys. All who do not have the right to try a case in court have not passed the bar. Therefore, all attorneys have a right to try a case in court. B, A, R

7. All internists are medical doctors. Jones is a philosopher and an internist. Therefore, it is not true that if Jones is a philosopher that he is not a medical doctor. I, M, j, P.

8. One is a utilitarian only if he believes in the greatest good for the greatest number is the criteria for ethical judgment. All non teleologists do not believe in the greatest good for the greatest number. No teleologists are deontologists. Therefore, no utilitarian is a deontologist. U, G, T, D.

9. All criminals should be prosecuted. One should be incarcerated only if he is guilty. Smith is a criminal and should be incarcerated. Therefore, it isn't true that Smith either should not be prosecuted or is not guilty.
C, P, I, G, s.

10. No non physicians are gynecologists. All who have not passed the bar exam are not attorneys. Smith is either a gynecologist or an attorney. Therefore, if Smith is not a physician, he has passed the bar exam.
P, G, B, A, s.

CHAPTER 32 FOUR NEW
BI-CONDITIONALS-LOGICAL
EQUIVALENCE

A. (x)(φx →ψx) ↔ ~(∃x)(φx & ~ψx)

If one reflects for a little bit on the ordinary English statements, "Everything in the lake is wet." and "There isn't anything in the lake which is not wet.", it becomes clear that they are logically equivalent. Thus we may say that "For any (x), if x is in the lake, then x is wet," (x)(Lx → Wx), is equivalent to "it is not the case that there exists an x such that x is in the lake and x is not wet," ~(∃x)(Lx & ~ Wx).

We can also demonstrate that this is the case by using previously learned rules of inference and replacement as is demonstrated below. In figure 32.1.

Figure 32.1

1.	(x)(Lx →Wx)	
2.	La →Wa	1 UI
3.	~La V Wa	2 IMP
4.	~(La & ~Wa)	3 DeM
5.	~(∃x)(Lx & ~ Wx)	5 EG

Having demonstrated this to be the case, we may posit a new rule of replacement. (Let us use the Greek letters φ (phi) and ψ (psi) to designate any predicates whatever. We may assert (x) (φx→ψx) ↔ ~(∃x) (φx & ~Ψx). In English this would read as follows: For any x, if φ may be predicated of x, then ψ may be predicated of x) is a bi-conditional of (it is not the case that there exists an x such that, φ may be predicated of x and not ψ may be predicated of x).

As we might expect the truth of this bi-conditional can also be demonstrated by a truth table as shown in figure 32.2.

Figure 32.2

φx	ψx	(x)(φ x	→	ψx)	↔	~(∃x)	(φx	&	~	ψx)
1	1	1	1	1	1	1	1	0	0	1
1	0	1	0	0	1	0	1	1	1	0
0	1	0	1	1	1	1	0	0	0	1
0	0	0	1	0	1	1	0	0	1	0

This relationship can also be illustrated by a flow chart and an algorithm as is shown below.

Figure 32.3

A: Flow Chart For (x)(φx → ψx) ↔ ~(∃x)(φx & ~ψx)

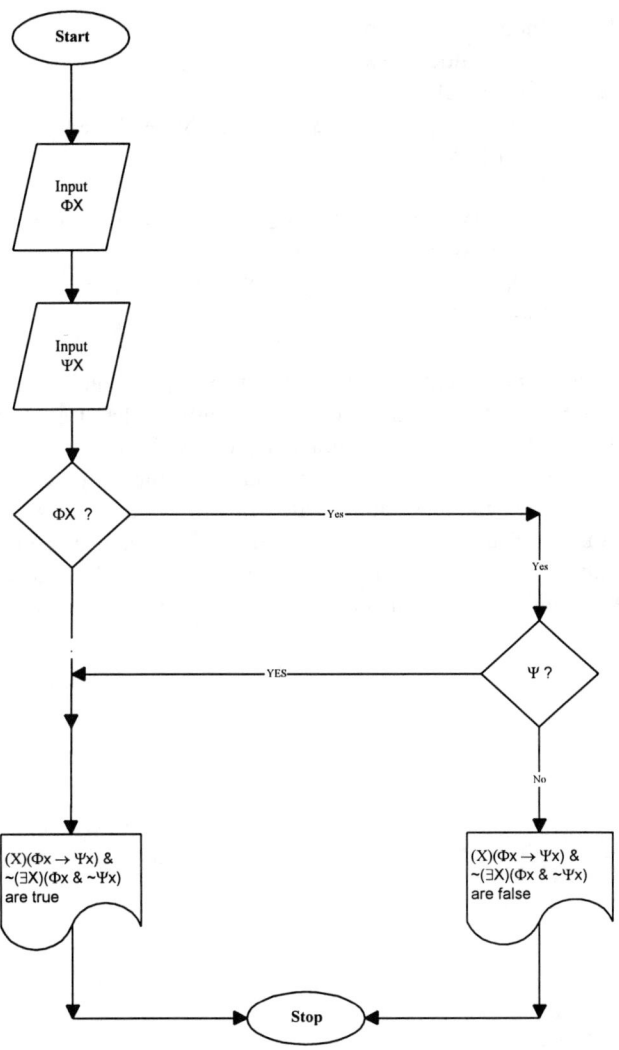

Figure 32.4

Algorithm For $(x)(\varphi x \to \psi x) \leftrightarrow \sim(\exists x)(\varphi x \mathbin{\&} \sim\psi x)$

1. Input value of φx.
2. Input value of ψx.
3. If φx is false,
   (a) $(x)(\varphi x \to \psi x) \mathbin{\&} \sim(\exists x)(\varphi x \mathbin{\&} \sim\psi x)$ are true.
   (b) Stop.
4. If ψx is true,
   (a) $(x)(\varphi x \to \psi x) \mathbin{\&} \sim(\exists x)(\varphi x \mathbin{\&} \sim\psi x)$ are true.
   (b) Stop.
5. $(x)(\varphi x \to \psi x) \mathbin{\&} \sim(\exists x)(\varphi x \mathbin{\&} \sim\psi x)$ are false.
6. Stop.

We have now demonstrated the logical equivalence of these two expressions by reflection in ordinary English, the use of rules of inference and replacement, a truth table, a flow chart, and an algorithm.

We may also develop some mechanical rules for changing any universal into an existential statement or vice a versa. These mechanical rules will work for **all four** of the logical equivalences introduced in this chapter. The rules will be illustrated below, however, with logical equivalency A which has been introduced above.

We will start with $(x)(\varphi x \rightarrow \psi x)$ and transform it into $\sim(\exists x(\varphi x \ \& \sim\psi x)$.

1. Change the sign in front of the quantifier from a positive to a negative or a negative to a positive.

$$\sim(x)(\varphi \ x \rightarrow \psi x)$$

2. Change the quantifier from a universal to an existential or an existential to a universal.

$$\sim(\exists \ x)(\varphi x \rightarrow \psi x)$$

3. Change the sign relating the two variables from an entailment to a conjunction or a conjunction to an entailment.

$$\sim(\exists \ x)(\varphi x \ \& \ \psi x)$$

4. Change the sign in front of the second variable from a positive to a negative or a negative to a positive.

$$\sim(\exists x)(\varphi x \ \& \sim\psi x)$$

B. $(\exists x)(\varphi x \ \& \ \psi x) \leftrightarrow \sim(x)(\varphi x \rightarrow \sim\psi x)$

Our second bi-conditional can also be observed clearly upon some reflection. If there exists an x such that both φ and ψ may be predicated of x; then it is not true that for any x, if φ may be predicated of x then $\sim\psi$ may be predicated of x.

This may perhaps be seen more clearly if we use a particular example. "There exists at least one individual who is both a native of Boston and of Irish descent or $(\exists \ x)(Bx \ \& \ Ix)$" is equivalent to saying that, "It's not true that no natives of Boston are of Irish descent, or $\sim(x)(Bx \rightarrow \sim Ix)$."

As in the previous case, this can be demonstrated by previously learned rules of inference and replacement as shown in figure 32.5.

Figure 32.5

1.	(∃x)(Bx & Ix)		
2.	Ba & Ia	1,	EI
4.	~(~Ba V ~Ia)	2,	DeM
5.	~(Ba → ~Ia)	3,	IMP
6.	~(x)(Bx → ~Ix)	4,	UG

Figure 32.6

Truth Table For (∃x) (φx & ψx) ↔ ~(x)(φx → ~ψx)

φx	ψx	(φx	&	ψx)	↔	~	(x)	(φx	→	~	ψx)
1	1	1	1	1	1	1		1	0	0	1
1	0	1	0	0	1	0		1	1	1	0
0	0	0	0	1	1	0		0	1	0	1
0	0	0	0	0	1	0		0	1	1	0

Figure 32.7

Flow Chart For $(\exists x)\,(\varphi x\ \&\ \psi x) \leftrightarrow \sim(x)(\varphi x \rightarrow \sim\Psi x)$

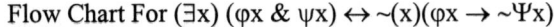

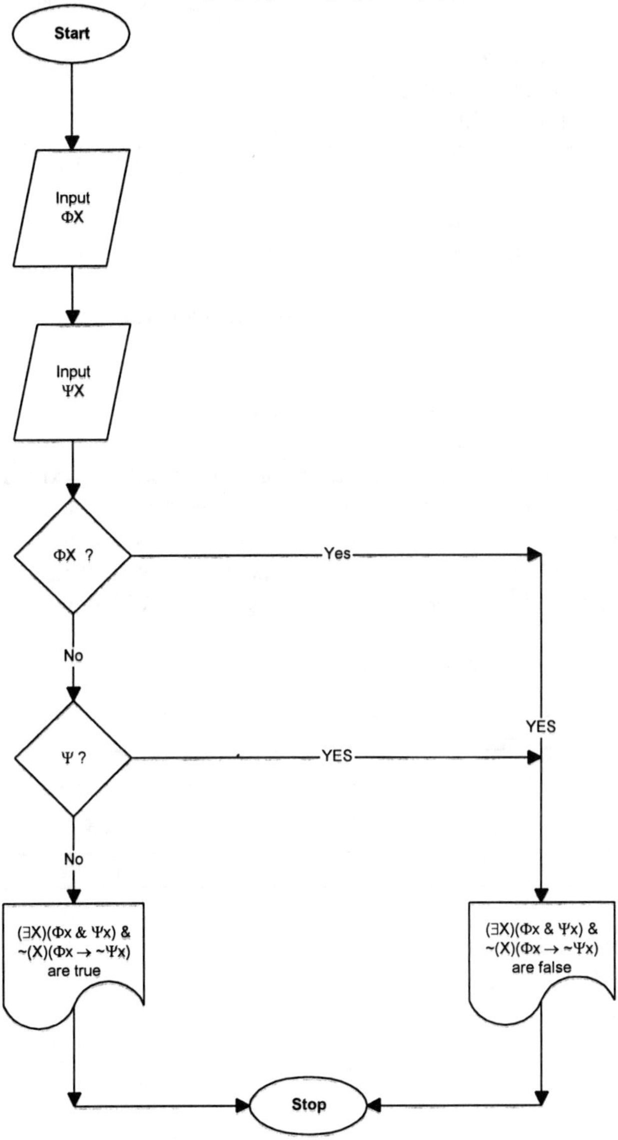

```
                    ┌──────────┐
                   (   Start    )
                    └──────────┘
                         │
                         ▼
                     ╱────────╲
                    │  Input   │
                    │   ΦX     │
                     ╲────────╱
                         │
                         ▼
                     ╱────────╲
                    │  Input   │
                    │   ΨX     │
                     ╲────────╱
                         │
                         ▼
                      ◇──────◇
                     ◇  ΦX ?  ◇──────────Yes──────────────────┐
                      ◇──────◇                                │
                         │                                    │
                        No                                   YES
                         │                                    │
                         ▼                                    │
                      ◇──────◇                                │
                     ◇   Ψ ?  ◇──────────YES──────────────┐   │
                      ◇──────◇                            ▼   ▼
                         │                                    │
                        No                                    ▼
                         │                           ┌──────────────────┐
                         ▼                           │ (∃X)(Φx & Ψx) &  │
            ┌──────────────────┐                     │ ~(X)(Φx → ~Ψx)   │
            │ (∃X)(Φx & Ψx) &  │                     │    are false     │
            │ ~(X)(Φx → ~Ψx)   │                     └──────────────────┘
            │    are true      │                              │
            └──────────────────┘                              │
                         │                                    │
                         ▼              ┌──────────┐          ▼
                         └─────────────(    Stop    )◄────────┘
                                        └──────────┘
```

Figure 32.8

Algorithm For (∃x) (φx & ψx) ↔ ~(x) (φx → ~ψx)

1. Input value of φx.
2. Input value of ψx.
3. If φx is false,
    (a) (∃x)(φx & ψx) & ~(x)(φx → ~ψ's) are false.
    (b) Stop.
4. If ψx is false,
    (a) (∃x)(φx & ψx) &~(x)(φx → ~ψx) are false.
    (b) Stop.
5. (∃x)(φ x & ψx) &~(x)(φx → ~ψx) are true.
6. Stop.

C. (x) (φx → ~ψx) → ~(∃x)(φx & ψx)

Demonstration of our third bi-conditional will be explained in a manner similar to the previous two. "No residents of Boston are Irish or (x)(Bx → ~Ix)" is a bi-conditional of "It's not true that some residents of Boston are Irish or ~(∃x)(Bx & Ix)." This is logically demonstrated in various ways in figures 32.9-32-12 below.

Figure 32.9

1.	(x)(Bx → ~Ix)		
2.	Ba → ~Ia	1,	UI
3.	~Ba V ~Ia	2,	IMP
4.	~(Ba & Ia)	3,	DeM
5.	~(∃x)(Bx & Ix)	4,	EG

Figure 32.10

Truth Tables For $(x)(\varphi x \rightarrow \sim\Psi x) \leftrightarrow \sim(\exists x)(\varphi x \ \& \ \Psi x)$

φx	Ψx	(φx)	→	~	(Ψx)	↔	~	(∃x)	(φx	&	Ψx)
1	1	1	0	0	1	1	0		1	1	1
1	0	1	1	1	0	1	1		1	0	0
0	1	0	1	0	1	1	1		0	0	1
0	0	0	1	1	0	1	1		0	0	0

Figure 32.11

Flow Chart For (x) (φx → ~Ψx) ↔ ~(∃x) (φx & Ψx)

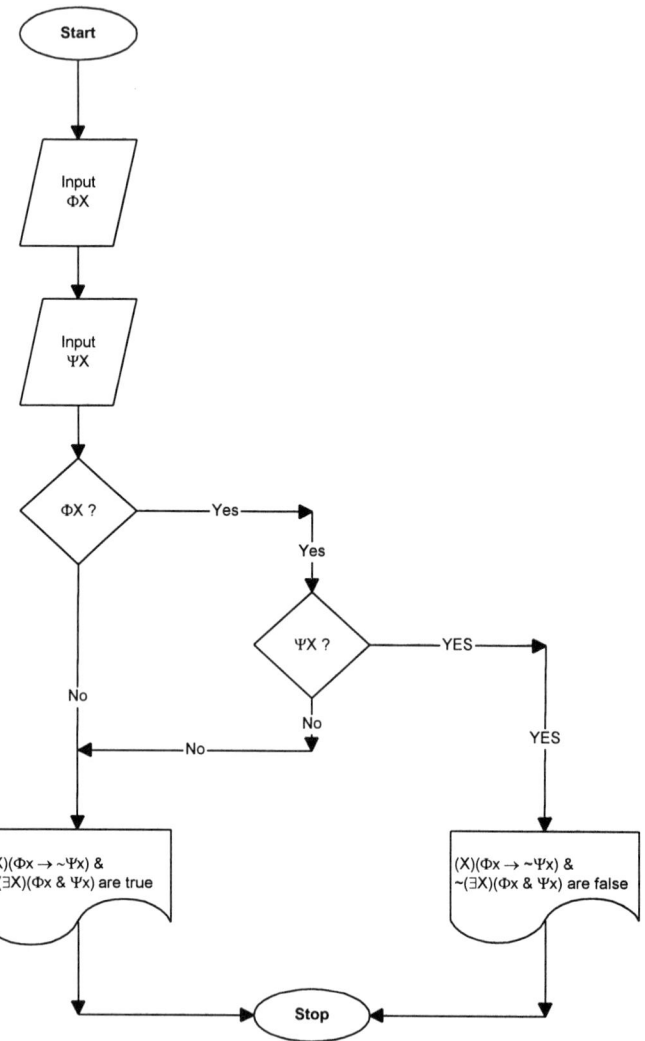

Figure 32.12

Algorithm For (x) (φx → ~Ψx) ↔ ~(∃ x) (φ x & Ψx)

1. Input value of φx.
2. Input value of Ψx.
3. If φx is false,
   (a) (x)(φx → ~Ψx) & ~(∃x)(φ x & Ψx) are true.
   (b) Stop.
4. If Ψx is false,
   (a) (x)(φ x → ~Ψx) & ~(∃x)(φx & Ψx) are true.
   (b) Stop.
5. (x)(φx → ~Ψx) & ~(∃x)(φx & Ψx) are false.
6. Stop.

D. (∃x)(φx & ~Ψx) ↔ ~(x)(Fφ → Ψx)

Our final bi-conditional will be explained as the previous ones. "Some residents of Boston are not Irish or (∃x)(Bx & ~Ix)" is a bi-conditional of "not all residents of Boston are Irish or ~(x)(φx → Ψx)" as is depicted in figures 32.13-32.16.

Figure 32.13

1.	(∃x)(φx & ~Ψx)		
2.	φx & ~Ψa	1,	EI.
3.	~(φx V ~Ψa)	2,	DeM.
4.	~(φx → ~Ψa)	3,	IMPL.
5.	~(x)(φx → ~Ψa))	4,	UG.

Logic of the Computing Sciences

Figure 32.14

Truth Table For (∃x) (φx & ~ Ψx) ↔ ~(x) (φx → Ψx)

φx	Ψx	∃x	(φx	&	~	Ψx)	↔	~	(x)	(φx	→	(Ψx)
1	1		1	0	0	1	1	0		1	1	1
1	0		1	1	1	0	1	1		1	0	0
0	1		0	0	0	1	1	0		0	1	1
0	0		0	0	1	0	1	0		0	1	0

Figure 32.15

Flow Chart For $(\exists x)\,(\varphi x\ \&\ \sim\!\Psi x)\leftrightarrow\ \sim\!(x)(\varphi x\rightarrow\Psi x)$

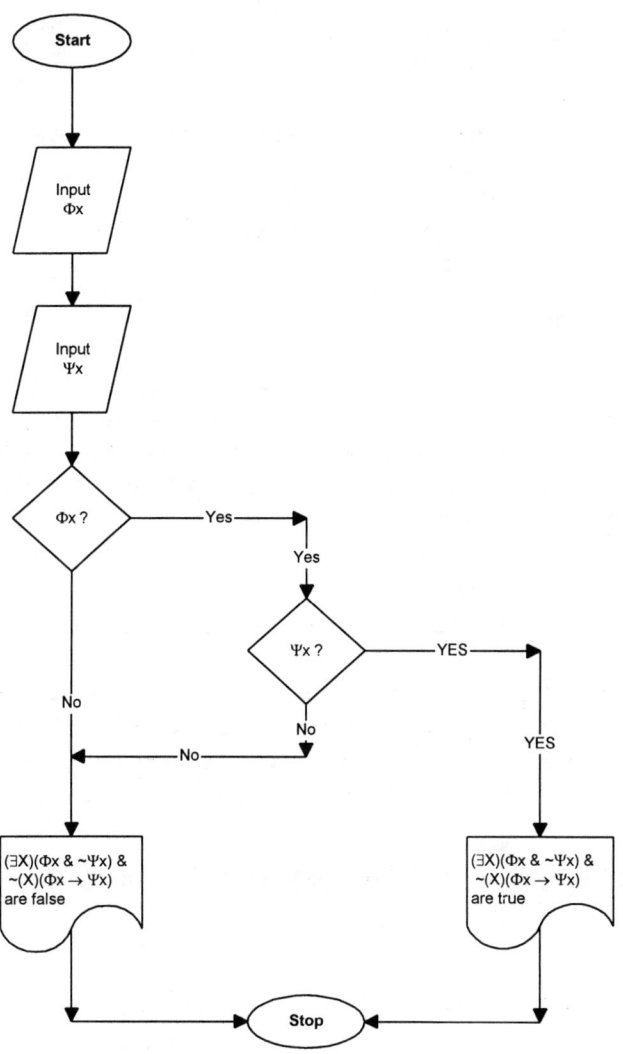

Figure 32.16

Algorithm For $(\exists x)(\varphi x \ \& \sim\Psi x) \leftrightarrow \sim(x)(\varphi x \rightarrow \Psi x)$

1. Input values of φx.
2. Input values of Ψx.
3. If φx is false,
    (a) $(\exists x)(\varphi x \ \& \sim\Psi x) \ \& \sim(x)(\varphi x \rightarrow \Psi x)$ are false.
    (b) Stop.
4. If Ψx is true,
    (a) $(\exists x)(\varphi x \ \& \sim\Psi x) \ \& \sim(X)(\varphi x \rightarrow \Psi x)$ are false.
    (b) Stop
5. $(\exists x)(\varphi x \ \& \sim\Psi x) \ \& \sim(X)(\varphi x \rightarrow \Psi x)$ are true.
6. Stop.

Since these relationships are logically equivalent, they can be treated as rules of replacement for purposes of solving problems. Rather than giving each logical equivalence a separate name we will designate all of these simply as Logical Equivalence and abbreviate them as LE. as in figure 32.17 below:

Figure 32.17

1.	$\sim(\exists x)(Ba \ \& \ Ix)$		$\therefore \sim Ia$
2.	Ba		
3.	$(x)(Bx \rightarrow \sim Ix)$	1,	LE.
4.	$Ba \rightarrow \sim Ia$	3	UI.
5.	$\sim Ia$	4,2	MP.

PROBLEM SET: FOUR BI-CONDITIONAL LOGICAL
EQUIVALENCES

1.
    1. $\sim(x)(Dx \rightarrow Fx)$ $\therefore (\exists x) \sim Fx$

2.
    1. $\sim(\exists x)(Fx \ \& \sim Cx)$ $\therefore \sim Fa$
    2. $\sim Ca$

3.

   1.   ~(x)(Hx → ~Ix)   ∴ ~Ja

   2.   (x)[Jx → (~Hx V ~Ix)

4.

   1.   ~(∃x)(Dx & ~Hx)   ∴ Ha

   2.   ~(x)(Bx → ~Dx)

5.

   1.   ~(∃x)(Hx & ~Fx)   ∴ ~Ha

   2.   ~(∃x)(Fx & Gx)

   3.   Ga

6.

   1.   ~(x)(Bx → Cx)   ∴~Sa

   2.   ~(∃x)(Sx & ~Cx)

7.

   1.   ~(∃x)(Bx & Cx)   ∴ ~(∃x)(Bx & Dx)

   2.   ~(∃x)(~Cx & Dx)

8.

   1.   ~(∃x)(~Ex & Fx)   ∴ ~(Ca & Fa)

   2.   ~(∃x)(Cx & Ex)

9.

   1.   ~(∃x)(Bx & ~Cx)   ∴ ~(∃x)(Bx & ~Ex)

   2.   ~(∃x)(~Ex & Dx)

   3.   (x)(Cx → Dx)

10.

   1.   ~(∃x)(~Hx & ~Fx)   ∴ Ha V (Fa & Ga)

   2.   ~(∃x)(~Hx & ~Gx)

1.   Its not the case that all attorneys are rich. Therefore, some attorneys who are not rich. A, R

2.   Jones is an attorney. Jones is also rich. Therefore, some attorneys are rich. j, A, R

3.   Jones is an attorney, Jones is also rich. Therefore, its not true that all attorneys are not rich. j, A, R

4.   It isn't true that there exists any attorneys who are rich. Smith is an
     attorney. Therefore, Smith is not rich.   A, R, s

5.   There doesn't exist an attorney who is not rich. All trial lawyers are
     attorneys. Jones is a trial lawyer. Therefore, Jones is rich.
                         A, R, T, j

6.   Its not true that no utilitarians are politically conservatives.   It isn't true
     that there are political conservatives who aren't readers of Burke.
     Therefore, its false that all utilitarians aren't readers of Burke.
                    Let U = utilitarians
                    Let P = political conservatives
                    Let B = readers of Burke

7.   Its not true that some Sunnites are not Muslims.   Its false that some
     Muslims are not monotheists. Only polytheists are believers in Zeus and
     Hera. There aren't any polytheists who are monotheists. Therefore, its
     erroneous to assert that there exist Sunnites who are believers in Zeus and
     Hera.
                    Let Sx = x is a Sunnite.
                    Let Mx = x is a Muslin.
                    Let Tx = x is a monotheist.
                    Let Px = x is a polytheist.
                    Let Zx = x is a believer in Zeus and Hera.

8. Its not true that there exists someone who is both a process theologian and a believer that God is not a completely temporal being. There isn't anyone who is both a believer that God perfectly knows the future and a believer that he is a completely temporal being, and also there isn't anyone who is not a believer that God completely controls the future and a believer that He is omnipotent. Furthermore, anyone who is a believer that God does not completely know the future is not a believer that God completely controls the future. Smith is a process theologian. Therefore, Smith is not a believer that God is omnipotent.   P, T, K, C, O, s

    Let Px = x is a process theologian.
    Let Tx = x is a believer that God is completely temporal.
    Let Kx = x is a believer that God perfectly knows the future.
    Let Cx = x is a believer that God completely controls the future.
    Let Ox = x is a believer that God is omnipotent.
    Let Os = Smith is a believer that God is omnipotent.

9. Its false that no Republicans are not social Darwinists. Its not true that there exists a Spencerian who is not a social Darwinist. Therefore, some Republicans are not Spencerians.   R, D, S

10. Its false that no democrats are true socialists. Its not true that there exists a true socialist who is not anti capitalist. Therefore, its not the case that no democrats are anti capitalists.   D, S, A

Study Guide

CHAPTER 4: ORDINARY ENGLISH AND "IF-THEN" STATEMENTS

1. $P \to E$ 2, $E \to P$ 3. $P \to E$ 4. $P \to E$ 5. $E \to P$
6. $E \to P$ 7. $E \to P$ 8. $E \to P$ 9. $P \to E$ 10. $E \to P$
11. $P \to H$ 12. $(E \to H) \to P$ 13. $(E \to\_P) \to H$
14. $(H \to E) \to P$ 15. $(P \to H) \to E$

CHAPTER 8: MODUS PONENS

3.
1. $F \to X$ $\therefore$ X
2. $C \to Y$
3. $Y \to R$
4. C
5. $R \to\_Z$
6. $Z \to\_F$
7. Y   2,4   M.P.
8. R   3,7   M.P.
9. Z   5,8   M.P.
10. F   6,9   M.P.
11. X   1,10 M.P.

5.
1. $[(D \to F) \to G]$ $\therefore$ $H \to (I \to J)$
2. $[(D \to F) \to G] \to [H \to (I \to J)]$
3. $H \to (I \to J)$ 2,1   M.P.

7.
1. $[(A \to B) \to (C \to D)] \to \{(D \to E) \to [(F \to G) \to (H \to I)]\}$ $\therefore$ I
2. H
3. $F \to G$
4. $D \to E$
5. J
6. $J \to [(A \to B) \to (C \to D)]$
7. $(A \to B) \to (C \to D)$ 6,5   M.P.
8. $(D \to E) \to [(F \to G) \to (H \to I)]$ 1,7   M.P.
9. $(F \to G) \to (H \to I)$ 8,4   M.P.
10. $H \to I$ 9,3   M.P.
11. I   10,2 M.P.

MODUS PONENS WORD PROBLEMS

3.
1. $L \to N$ $\therefore$ N
2. L
3. N   1,2 M.P.

5.
1. $(N \to G) \to (C \to P)$ $\therefore$ P
2. $N \to G$
3. C
4. $C \to P$ 1,2 M.P.
5. P   4,3 M.P.

7.
1. $G \to (N \to S)$ $\therefore$ S
2. N
3. G

    4. N → S   1,3 M.P.
    5. S   4,2 M.P.

CHAPTER 9: MODUS TOLLENS

3. 1. [(A → B) → (D → C)] → D   ∴ ~[(A → B) → (D → C)]
    2. ~D
    3. ~[(A → B) → (D → C)]   1,2 M.T.

5. 1. ~C   ∴ H
    2. (A → B) → C
    3. ~(A → B) → [(D → F) → (G → H)]
    4. G
    5. D → F
    6. ~(A → B)   2,1 M.T.
    7. (D → F) → (G → H)   3,6 M.P.
    8. G → H   7,5 M.P.
    9. H   8,4 M.P.

7. 1. (A → B) → [(C → D) → (E → F)]   ∴ F
    2. ~(G → F) → (A → B)
    3. (G → F) → I
    4. ~I
    5. ~I → (C → D)
    6. E
    7. ~(G → F)   3,4 M.T.
    8. A → B   2,7 M.P.
    9. (C → D) → (E → F)   1,8 M.P.
    10. C → D   5,4 M.P.
    11. E → F   9,10 M.P.
    12. F   11,6 M.P.

MODUS TOLLENS WORD PROBLEMS

3. 1. C → ~P   ∴ ~S
    2. C
    3. S → P
    4. ~P   1,2 M.P.
    5. ~S   3,4 M.T.

6. 1. P → C   ∴ ~_P
    2. C → L
    3. L → U
    4. ~U
    5. ~L   3,4 M.T.
    6. ~C   2,5 M.T.
    7. ~P   1,6 M.T.

7. 1. P → ~C   ∴ ~P
    2. ~C → ~L
    3. ~L → ~U

4. ~~U
5. ~~L   3,4 M.T.
6. ~~C   2,5 M.T.
7. ~P   1,6 M.T.

CHAPTER 10: CONJUNCTION AND SIMPLIFICATION

3. 1. [~(O → P) & (T → W)] → [(H & W) → (R & S)]   ∴ R
    2. (T → W) & (A & B)
    3. ~(R → S)
    4. (O → P) → (R → S)
    5. H & W
    6. ~(O → P)   4,3 M.T.
    7. T → W   2   SIMP.
    8. ~(O → P) & (T → W)   6,7 CONJ.
    9. (H & W) → (R & S)   1,8 M.P.
   10. R & S   9,5 M.P.
   11. R   10 SIMP.

5. 1. {[(S → R) & (P → R)] & (Q → R)]} → [(R → S) & (R → Q)]   ∴ R → S
    2. Q → R
    3. (S → R) &(P → R)
    4. [(S → R) &(P → R)] & (Q → R)   3,2   CONJ.
    5. (R → S) & (R → Q)   1,4   M.P.
    6. R → S   5,   SIMP.

7. 1. [~(P → Q)&~(R → S)] → (~A & ~T)   ∴ ~A
    2. (P → Q) → (C → D)
    3. ~(C → D)
    4. (R → S) → (C → D)
    5. ~(P → Q)   2,3 M.T.
    6. ~(R → S)   4,3 M.T.
    7. ~(P → Q) & ~(R → S)   5,6 CONJ.
    8. ~A & ~T   1,7 M.P.
    9. ~A   8 SIMP.

CONJUNCTION AND SIMPLIFICATION WORD PROBLEMS

3. 1. (W & ~S) → (O & T)   ∴ B
    2. ~(O & T)
    3. ~(W & ~S) → B
    4. ~(W & ~S)   1,2 M.T.
    5. B   3,4 M.P.

5. 1. P → E$_1$ ∴ I
    2. A → E$_2$
    3. E$_1$ → S
    4. S → (~K & ~O)
    5. ~(~K & ~O)
    6. ~P → A
    7. E$_2$ → (D & N)
    8. D → I

9. ~S   4,5 M.T.
10. ~E₁   3,9 M.T.
11. ~P   1,10 M.T.
12. A   6,11 M.P.
13. E₂   2,12 M.P.
14. D & N   7,13 M.P.
15. D   14 SIMP.
16. I   8,15 M.P.

7. 1. B → R   ∴ A
   2. F → L
   3. F
   4. (R & L) → A
   5. B
   6. R   1,5 M.P.
   7. L   2,3 M.P.
   8. R & L   6,7 CONJ.
   9. A   4,8 M.P.

CHAPTER 11: HYPOTHETICAL SYLLOGISM

3. 1. [(A → B) & (C → D)] → [(E → F) & (G → H)]
   2. [(E → F) & (G → H)] → [(I → J) & (K → L)]
   3. ~[(I → J) & (K → L)]
   4. [(A → B) & (C → D)] → [(I → J) & (K → L)] 1,2 H.S.
   5. ~[(A → B) & (C → D)]   4,3 M.T.
                                ∴ ~[(A → B) & (C → D)]

5. 1. {[(A → B) & (C → D)] → [(E → F) → (I → J)]} →_{[(K → L) &
       (M → N)] → (O → P)}
   2. Q → {[(A → B) & (C → D)] → [(E → F) → (I → J)]}
   3. {[(K → L) & (M → N)] → (O → P)} → R
   4. R → S   ∴ Q → S
   5. Q → {[(K → L) & (M → N)] → (O → P)} 2,1 H.S.
   6. Q → R   5,3 H.S.
   7. Q → S   6,4 H.S.

7. 1. [(H & I) & (J & K)] → {[(L & M) & (N & O)] & (P & Q)}   ∴ A → Z
   2. [(E & F) & G] → [(H & I) & (J & K)]
   3. (C & D) → [(E & F) & G]
   4. {[(L & M) & (N & O)] & (P → Q)} → {[(R → S) → (T → U)] & (V → W)}
   5. B → (C & D)
   6. {[(R → S) → (T → U)] & (V → W)} → (X → Y)
   7. A → B
   8. (X → Y) → Z
   9. A → (C & D)   7,5 H.S.
   10. A → [(E & F) & G]   9,3 H.S.
   11. A → [(H & I) & (J & K)]   10,2 H.S.
   12. A → {[(L & M) & (N & O)] & (P & Q)} 11,1 H.S.
   13. A → {[(R → S) → (T → U)] & (V → W)} 12,4 H.S.
   14. A → (X → Y)   13,6 H.S.

15. A → Z   14,8 H.S.

HYPOTHETICAL SYLLOGISM WORD PROBLEMS

3. 1. O → I   ∴ ~O
    2. I → D3
    3. D3 → M
    4. M → A
    5. ~A
    6. O → D3   1,2 H.S.
    7. O → M   6,3 H.S.
    8. O → A   7,4 H.S.
    9. ~O   8,5 M.T.

5. 1. G → D   ∴ L
    2. D → (B & P)
    3. (A & B) → L
    4. G
    5. A
    6. G → (B & P)   1,2 H.S.
    7. B & P   6,4 M.P.
    8. B   7   SIMP.
    9. A & B   5,8 CONJ.
    10. L   3,9 M.P.

7. 1. (B & D) → (A → L)   ∴ P → R
    2. P → (B & D)
    3. (A → L) → R
    4. P → (A → L)   2,1 H.S.
    5. P → R   4,3 H.S.

CHAPTER 12 : ABSORPTION

3. 1. D → C   ∴ ~D
    2. C → R
    3. (C & R) → S
    4. S → T
    5. (S & T) → U
    6. ~[(S & T) & U]
    7. (S & T) → [(S & T) & U]   5   ABS.
    8. ~(S & T)   7,6   M.T.
    9. S → (S & T)   4   ABS.
    10. ~S   9,8   M.T.
    11. ~(C & R)   3,10   M.T.
    12. C → (C & R)   2   ABS.
    13. ~C   12,11 M.T.
    14. ~D   1,13   M.T.

5. 1. D → C   ∴ ~C
    2. C → R
    3. (C & R) → S
    4. S → T
    5. (S & T) → U

  6. ~[(S & T) & U]
  7. (S & T) → [(S & T) & U]   5   ABS.
  8. ~(S & T)   7,6 M.T.
  9. S → (S & T)   4   ABS.
10. ~S   9,8 M.T.
11. ~(C & R)   3,10   M.T.
12. C → (C & R)   2   ABS.
13. ~C   12,11 M.T.

7. 1. C → D   ∴ C → C & {D & [E & (F & G)]}
    2. D → E
    3. E → F
    4. F → G
    5. F → (F & G)   4, ABS
    6. E → (F & G)   3,5 H.S.
    7. E → [E & (F & G)]   6, ABS
    8. D → [E & (F & G)]   2,7 H.S.
    9. D → {D & [E & (F & G)]} 8, ABS
10. C → {D & [E & (F & G)]} 1,9 H.S.
11. C → C & {D & [E & (F & G)]} 10, ABS

ABSORPTION WORD PROBLEMS

3. 1. D1 → S1   ∴ ~D
    2. S1 → S2
    3. S2 → S3
    4. ~[S1 & (S2 & S3)]
    5. S2 → (S2 & S3)   3 ABS
    6. S1 → (S2 & S3)   2,5 H.S.
    7. S1 → [S1 & (S2 & S3)]   6 ABS
    8. D1 → [S1 & (S2 & S3)]   1,7 H.S.
    9. ~D1   8,4 M.T.

5. 1. S → H   ∴ S → (T & I)
    2. H → U
    3. (H & U) → P
    4. P → T
    5. T → I
    6. T → (T & I)   5 ABS.
    7. P → (T & I)   4,6 H.S.
    8. (H & U) →_ (T & I)   3,7 H.S.
    9. H → (H & U)   2, ABS.
10. H → (T & I)   9,8 H.S.
11. S → (T & I)   1,10 H.S

7. 1. P → D   ∴ ~S
    2. (P & D) → ~A
    3. P
    4. S → A
    5. P → (P & D)   1 ABS.
    6. P & D   5,3 M.P.

7. ~A   2,6 M.P.
8. ~S   4,7 M.T.

CHAPTER 14: DISJUNCTIVE SYLLOGISM

3. 1. ~F   ∴ A
    2. F V [(A & B) & C]
    3. (A & B) & C   2,1 D.S.
    4. A & B   3,   SIMP.
    5. A   4,   SIMP.

5. 1. G → H   ∴ ~A
    2. G V [(A → D) & F]
    3. ~(G & H)
    4. ~D
    5. G → (G & H)   1   BS.
    6. ~G   5,3 M.T.
    7. (A → D) & F   2,6 D.S.
    8. A → D   7   SIMP.
    9. ~A   8,4 M.T.

7. 1. A V [(C & D) & (R → S)]   ∴ C
    2. (E & F) → (~G & H)
    3. F
    4. E
    5. A → G
    6. E & F   4,3 CONJ.
    7. (~G & H)   2,6 M.P.
    8. ~G   7   SIMP.
    9. ~A   5,8 M.T.
    10. (C & D) & (R V S)   1,9 D.S.
    11. C & D   10   SIMP.
    12. C   11   SIMP.

DISJUNCIVE SYLLOGISM WORD PROBLEMS

3. 1. (D → L) V A   ∴ A
    2. ~(D → L)
    3. A   1,2 D.S.

5. 1. (A → H) V (H → T)   ∴ (H → T)
    2. ~(A → H)
    3. H → T   1,2 D.S.

7. 1. B V C   ∴ C & A
    2. B → I
    3. C → A
    4. ~I
    5. ~B   2,4 M.T.
    6. C   1,5 D.S.
    7. A   3,6 M.P.
    8. C & A   6,7 CONJ.

CHAPTER 15: CONSTRUCTIVE DILEMMA

3. 1. A V B   ∴ (D → F) V (D → F)
    2. A → C
    3. C → (D → F)
    4. B → C
    5. A → (D → F)   2,3 H.S.
    6. B → (D → F)   4,3 H.S.
    7. [A → (D → F)] & [B → (D → F)] 5,6 CONJ.
    8. (D → F) V (D → F)   7,1 C.D.

5. 1. (F V G) & (H → I)   ∴ K V O
    2. (F → K) & (L → M)
    3. (G → 0) & (P → Q)
    4. F → K   2,   SIMP.
    5. G → O   3,   SIMP.
    6. (F → K) & (G → O)   4,5 CONJ.
    7. F V G   1,   SIMP.
    8. K V O   6,7 C.D.

7. 1. R → (S V T)   ∴ F V Z
   2   .U → (W V X)
   3. (~R & ~U) → T V Y
   4. T → F
   5. ~(S V T)
   6. Y → Z
   7. ~(W V X)
   8. ~R   1,5 M.T.
   9. ~U   2,7 M.T.
   10. ~R & ~U   8,9 CONJ.
   11. T V Y   3,10 M.P.
   12. (T → F) & (Y → Z)   4,6 CONJ.
   13. F V Z   12,11   C.D.

CONSTRUCTIVE DILEMMA WORD PROBLEMS

3. 1. S → F   ∴ F V E
    2. M → E
    3. S V M
    4. (S → F) & (M → E)   1,2 CONJ
    5. F V E   4,3 C.D.

5. 1. A → B   ∴ B V C
    2. D → C
    3. A V D
    4. (A → B) & (D → C)   1,2 CONJ
    5. B V C   4,3 C.D.

7. 1. S → E   ∴ (E & I) V (C & J)
    2. E → I
    3. ~S → C
    4. C → J

5. S V ~S
6. E → (E & I)   2,   ABS.
7. C → (C & J)   4,   ABS.
8. S → (E & I)   1,6   H.S.
9. ~S → (C & J)   3,7   H.S.
10. [S → (E & I)] & [~S → (C & J)]   8,9   CONJ.
11. (E & I) V (C & J)   10,5 C.D

CHAPTER 16: ADDITION

3. 1. C V E   ∴ (F V G) V {(P & B) → [D → (R V S)]}
    2. C → D
    3. E → (F V G)
    4. ~D
    5. ~C   2,4 M.T.
    6. E   1,5 D.S.
    7. F V G   3,6 M.P.
    8. (F V G) V [(P & B) → [D → (R V S)]}   7,   ADD.

5. 1. {[(A & B) → (C V D) & (E V F)} → G   ∴ G V H
    2. E V F
    3. (A & B) → (C V D)
    4. [(A & B) → (C V D)] & (E V F)   3,2 CONJ.
    5. G   1,4 M.P.
    6. G V H   5,   ADD.

7. 1. (A & B) → J   ∴ K V J
    2. (~C V D) → K
    3. ~C
    4. ~C V D   3,   ADD.
    5. (~C V D) V (A & B)   4,   ADD.
    6. [(~C V D) → K] & [(A & B) → J]   2,1 CONJ.
    7. K V J   6,5 C.D.

ADDITION WORD PROBLEMS

3. 1. (P V J) → M   ∴ M V N
    2. P
    3. P V J   2,   ADD.
    4. M   1,3 M.P.
    5. M V N   4,   ADD.

5. 1. M → T   ∴ (T V J) V (C & D)
    2. A → J
    3. M V A
    4. (M → T) & (A → J)   1,2 CONJ.
    5. T V J   4,3 C.D.
    6. (T V J) V (C & D)   5   ADD.

7. 1. U_1 → T_1 ∴ $(T_1$ V $~T_2)$ V $(T_3$ V $~T_3)$
    2. $~U_2$ → $~T_2$
    3. U_1 V $~U_2$
    4. $(U_1 → T_1)$ & $(~U_2 → ~T_2)$ 1,2 CONJ.
    5. T_1 V $~T_2$ 4,3 C.D.

6.   (T₁ V ~T₂) V (T₃ V ~T₃)   5,   ADD.

CHAPTER 21: DOUBLE NEGATION

3. 1. (~~ D _→ E) _→ F   ∴ ~~ F
    2. D _→ E̅
    3. ~~ D _→ E   2,   D.N.
    4. F   1,3   M.P.
    5. ~~ F   4   D.N.

5. 1. D _→ E   ∴ ~ D & ~~ ~ F
    2.  ~~~E
    3. F→ ~ G
    4. G
    5. ~ E   2,   D.N.
    6. ~ D   1,5   M.T.
    7. ~~ G   4,   D.N.
    8. ~ F   3,7   M.T.
    9. ~~~F   8   D.N.
    10. ~D & ~~~F   6,9   CONJ.

7. 1. H → (I & J)   ∴ ~~ H → [H & (I & ~~ J)]
    2. H → [H & (I & J)   1,   ABS.
    3. ~~ H → [H & (I & J)]   2,   D.N.
    4. ~~ H → [H & (I & ~~ J)]   3,   D.N.

DOUBLE NEGATION: WORD PROBLEMS

3. 1. R _→ A   ∴ ~ R & ~ A
    2. A _→ M
    3. ~ E̅ _→ ~ G
    4. M _→ G
    5. C _→ ~ E
    6. ~~ C
    7. C   6,   D.N.
    8. ~ E   5,7   M.P.
    9. ~ G   3,8   M.P.
    10. ~ M   4,9   M.T.
    11. ~ A   2,10 M.T.
    12. ~ R   1,11 M.P.
    13. ~ R & ~ A   12,11 CONJ.

5. 1. ~ I → ~ U   ∴ I
    2. ~ U → ~ J
    3. J
    4. ~ I → ~ J   1,2   H.S.
    5. ~~ J   3   D.N.
    6. ~~ I   4,5   M.T.
    7. I   6   D.N.

7. 1. ~(H V B) → (I V S)   ∴ (H V B) V C
    2. ~(I V S)

3. ~~(H V B)   1,2 M.T.
4. (H V B)   3   D.N.
5. (H V B) V C   4   ADD.

CHAPTER 22: COMMUTATION

3. 1. (A & B) _→ (C & D)   ∴ D & C
    2. B & A
    3. A & B   2,   COM.
    4. C & D   1,3 M.P.
    5. D & C   4,   COM.

5. 1. (A & B) → (C & D)   ∴ (B & A) & (D & ~~ C)
    2. B & A
    3. A & B   2 COM.
    4. (A & B) → [(A & B) & (C & D)] 1, ABS.
    5. (A & B) & (C & D)   4,3 M.P.
    6. (B & A) & (C & D)   5 COM.
    7. (B & A) & (D & C)   6 COM.
    8. (B & A) & (D & ~~ C)   7 D.N.

7. 1. D → E   ∴ G
    2. F → (G V H)
    3. F V D
    4. ~(D & E)
    5. ~ H
    6. D → (D & E)   1,   ABS.
    7. [F → (G V H) & [D → (D & E) 2,6 CONJ.
    8. (G V H) V (D & E)   7,3 C.D.
    9. (D & E) V (G V H)   8,   COM.
    10. G V H   9,4 D.S.
    11. H V G   10, COM.
    12. G   11,5 D.S.

COMMUTATION WORD PROBLEMS

3. 1. (H → L) & (M → T)   ∴ (M → T) & (H → L)
    2. (M → T) & (H → L)   1, COM.

5. 1. (H → L) & (T → M)   ∴ L V M
    2. T V H
    3. H V T   2 COM.
    4. L V M   1,3 C.D.

7. 1. [(U & O) V (D & F)] & [(I & P) V (S & M)]
    2. [(I & P) V (S & M)] & [(U & O) V (D & F)]   1,   COM.
    3. [(P & I) V (S & M)] & [(U & O) V (D & F)]   2,   COM.
    4. [(P & I) V (M & S)] & [(U & O) V (D & F)]   3,   COM.
    5. [(P & I) V (M & S)] & [(D & F) V (U & O)]   4,   COM.
                ∴ [(P & I) V (M & S)] & [(D & F) V (U & O)]

CHAPTER 23: TAUTOLOGY

3. 1. D   ∴ (D V D) & (E & E)
   2. E
   3. D V D   1,   TAUT.
   4. E & E   2,   TAUT.
   5. (D V D) & (E & E)   3,4 CONJ.

5. 1. [(A & A) V A] & [~~A V (A & A)]   ∴ A
   2. [(A & A) V A] & [A V (A & A)]   1,   D.N.
   3. [(A & A) V A] & [(A & A) V A]   2,   COM.
   4. (A & A) V A   3,   TAUT.
   5. A V A   4,   TAUT.
   6. A   5,   TAUT.

7. 1. D → R   ∴ ~~(~~R V X) V (C V R)
   2. E → C
   3. (D & D) V (E & E)
   4. D V (E & E)   3   TAUT.
   5. D V E   4   TAUT.
   6. (D → R) & (E → C)   1,2 CONJ.
   7. R V C   6,5 C.D.
   8. C V R   7   COM.
   9. (C V R) V~~(~~R V X)   8,   ADD.
   10. ~~(~~R V X) V (C V R)   9,   COM.

TAUTOLOGY WORD PROBLEMS

3. 1. P → W   ∴ W
   2. P & P
   3. P   2,   TAUT.
   4. W   1,3 M.P.

5. 1. P → C   ∴ C
   2. F → C
   3. P V F
   4. (P → C) & (F → C)   1,2 CONJ.
   5. C V C   4,3 C.D.
   6. C   5,   TAUT.

7. 1.{[(R → P) & (R → P)] V (R → P)} & {[(R → P) V (R → P)] V (R → P)}
   2. R
   3. [(R → P) V (R → P)] & {[(R _→ P) V (R → P)] V (R → P)} 1, TAUT.
   4. (R → P) & {[(R → P) V (R → P)] V (R → P)} 3, TAUT.
   5. (R → P) & [(R → P) V (R → P)]   4, TAUT.
   6. (R → P) & (R → P)   5, TAUT.
   7. R → P   6, TAUT.
   8. P   7,2 M.P.
                                              ∴ P

CHAPTER 24: ASSOCIATION

3. 1. (C V B) V A   ∴ (A V B) V C
   2. A V (C V B)   1,   COM.
   3. A V (B V C)   2,   COM.
   4. (A V B) V C)   3,   ASSOC.

5. 1. (A V D) V (B V C)   ∴ (A V B) V (C V D)
   2. [(A V D) V B] V C   1,   ASSOC.
   3. [A V (D V B)] V C   2,   ASSOC.
   4. [A V (B V D)] V C   3,   COM.
   5. [(A V B) V D] V C   4,   ASSOC.
   6. (A V B) V (D V C)   5,   ASSOC.
   7. (A V B) V (C V D)   6,   COM.

7. 1. (C V B)   ∴ (A V B) V C
   2. (C V B) V A   1   ADD.
   3. A V (C V B)   2   COM.
   4. A V (B V C)   3,   COM.
   5. (A V B) V C   4,   ASSOC.

ASSOCIATION WORD PROBLEMS

3. 1. I   ∴ [(C V F) V I] & [C V (F V I)]
   2. I V (C V F)   1   ADD.
   3. (C V F) V I   2   COM.
   4. C V (F V I)   3   ASSOC.
   5. [(C V F) V I] & [C V (F V I)]   3,4   CONJ.

5. 1. D & (R & H)   ∴ C
   2. (D & R) → C
   3. (D & R) & H   1,   ASSOC.
   4. D & R   3,   SIMP.
   5. C   2,4   M.P.

7. 1. R → (E V U)   ∴ (U V V) V (E V P)
   2. M → (V V P)
   3. M V R
   4. [M → (V V P)] & [R → (E V U)]   1,2   CONJ.
   5. (V V P) V (E V U)   4,3   C.D.
   6. [(V V P) V E)] V U   5   ASSOC.
   7. [V V (P V E)] V U   6   ASSOC.
   8. [V V (E V P)] V U   7   COM.
   9. U V [V V (E V P)]   8   COM.
   10. (U V V) V (E V P)   9   ASSOC.

CHAPTER 25: TRANSPOSITION

3. 1. ~ ~ C → D   ∴ ~ D → ~ C
   2. ~ D → ~ C   1,   TRANS.

5. 1. ~ A → ~ B   ∴ A V C
   2. ~ C → D
   3. B V ~ D

4. B → A	1,	TRANS.
5. ~ D → C	2,	TRANS.
6. (B → A) & (~ D → C)	4,5	CONJ.
7. A V C	6,3	C.D.

7. 1. ~ (~ B → ~ A) → ~ (~ D → ~ C)   ∴ B
    2. C → D
    3. A

4. (~ D → ~ C) → (~ B → ~ A)	1,	TRANS.
5. (C → D) → (~ B → ~ A)	4,	TRANS.
6. ~ B → ~ A	5,2	M.P.
7. A → B	6,	TRANS.
8. B	7,3	M.P.

WORD PROBLEMS TRANSPOSITION

3. 1. ~ [(E & F) & (D & M)] → L   ∴ ~ L → [(E & F) & (D & M)]
    2. ~ L → [(E & F) & (D & M)]   1, TRANS.

5. 1. (B → P) → H   ∴ ~H → ~(~P → ~B)
    2. ~H → ~(B → P)   1, TRANS.
    3. ~H → ~(~P → ~B)   2, TRANS.

7. 1. (W & A) → (T & N)   ∴ ~(T & N) → ~(W & A)
    2. ~(T & N) → ~(W & A)   1, TRANS.

CHAPTER 26: MATERIAL IMPLICATION.

3. 1. ~ A → B   ∴ A V B
    2. A V B   1, IMPL..

5. 1. A V B   ∴ ~ A → (~ A & B)
    2. ~ A → B   1, IMPL..
    3. ~ A → (~ A & B)   2, ABS.

7. 1. ~ D → ~ C   ∴ A → D
    2. ~ B → ~ A
    3. ~~ C V ~ B

4. ~ C → ~ B	3,	IMPL..
5. ~ D → ~ B	1,4	H.S.
6. ~ D → ~ A	5,2	H.S.
7. A → D	6,	TRANS.

WORD PROBLEMS MATERIAL IMPLICATION

3. 1. ~M V L   ∴ M → L
    2. M → L   1, IMPL.

5. 1. ~(D → H) → R   ∴ (D → H) V R
    2. (D → H) V R   1, IMPL.

7. 1. (H → G) V (H → M)   ∴ ~H V (M V G)
    2. (~H V G) V (H → M)   1,   IMPL.
    3. (~H V G) V (~H V M)   2   IMPL.
    4. ~H V [G V (~H V M)]   3   ASSOC.
    5. ~H V [(G V ~H) V M]   4   ASSOC.
    6. ~H V [(~H V G) V M]   5   COM.
    7. ~H V [~H V (G V M)]   6   ASSOC.
    8. ~H V [~H V (M V G)]   7   COM.
    9. (~H V ~H) V (M V G)   8   ASSOC.
   10. ~H V (M V G)   9   TAUT.

CHAPTER 27: EXPORTATION

3. 1. ~ F   ∴ ~ (D & E)
    2. E → (D → F)
    3. (E & D) → F   2   EXP.
    4. (D & E) → F   3   COM.
    5. ~(D & E)   4,1 M.T.

5. 1. (A & B) → [(C & D) → (E & F)]
    2. (A & B) → {C → [D → (E & F)]}   1, EXP.
    3. A → B → {C → [D → (E & F)]}   2, EXP.
                ∴ A → B → {C → [D → (E & F)]}

7. 1. A V (B V C)   ∴ (~ A & ~ B) → C
    2 ~A → (B V C)   1,   IMPL..
    3 ~A → (~B → C)   2,   IMPL..
    4 (~A & ~B) → C   3,   EXP.

EXPORTATION WORD PROBLEMS

3. 1. (O & L) → E   ∴ ~(O → E) → ~_L
    2. (L & O) → E   1,   COM.
    3. L → (O → E)   2,   EXP.
    4. ~(O → E) → ~L   3,   TRANS.

5. 1. (F V V) V R   ∴ (~ F & ~V) → R
    2. F V (V V R)   1,   ASSOC.
    3. ~F → (V V R)   2,   IMPL.
    4. ~F → (~V → R)   3,   IMPL.
    5. (~F & ~V) → R   4,   EXP.

7. 1. ~[F → (V → R)] → W   ∴ ~ R → ~ (F & V)
    2. ~ W
    3. ~~ [F → (V → R)]   1,2, M.T.
    4. F → (V → R)   3,   D.N.
    5. (F & V) → R   4,   EXP.
    6. ~ R → ~ (F & V)   5,   TRANS.

CHAPTER 28: MATERIAL EQUIVALENCE

3. 1. D ↔ C   ∴ ~ D
    2. ~ C
    3. (D → C) & (C → D)   1   EQUIV.
    4. D → C   3   SIMP.
    5. ~ D   4,2 M.T.

5. 1. ~ D   ∴ D ↔ C
    2. ~ C
    3. ~ D & ~ C   1,2, CONJ.
    4. (~ D & ~ C) V (D & C)   3,   ADD.
    5. (D & C) V (~ D & ~ C)   4,   COM.
    6. D ↔ C   5,   EQUIV.

7. 1. [~ A V (B V C)] & [~ B V (A V C)]   ∴ A ↔ B
    2. ~ C
    3. ~ A V (B V C)   1, SIMP.
    4. (~ A V B) V C   3, ASSOC.
    5. C V (~ A V B)   4, COM.
    6. ~ A V B   5,2 D.S.
    7. A → B   6, IMPL.
    8. [~ B V (A V D)] & [~ A V (B V C)] 1, COM.
    9. ~ B V (A V C)   8, SIMP.
    10. (~ B V A) V C   9, ASSOC.
    11. C V (~ B V A)   10, COM.
    12. ~ B V A   11,2, D.S.
    13. B → A   12, IMPL..
    14. (A → B) & (B → A)   7,13, CONJ.
    15. A ↔ B   14, EQUIV.

MATERIAL EQUIVALENCE WORD PROBLEMS

3. 1. F → (S & A)   ∴ F ↔ (S & A)
    2. A → (S → F)
    3. (A & S) → F   2   EXP.
    4. (S & A) → F   3, COM.
    5. [F → (S & A)] & [(S & A) → F]   1,4 CONJ.
    6. F ↔ (S & A)   5   EQUIV.

5. 1. F V P   ∴ ~ F ↔ P
    2. P → ~ F
    3. ~~ F V P   1   D.N.
    4. ~ F → P   3   IMP.
    5. (~ F → P) & (P → ~F)   4,2 CONJ.
    6. (~ F ↔ P)   5   EQUIV.

7. 1. ~P V L   ∴ P ↔ L
    2. ~P → ~L
    3. P → L   1, IMP.
    4. L → P   2, TRANS.
    5. (P → L) & (L → P)   3,4, CONJ.

6. P↔ L   5, EQUIV.

CHAPTER 29: DISTRIBUTION

3. 1.   E V (F & G)   ∴ A ↔ B
    2.   ~ F V (A & B)
    3.   (~ B & ~ A) V ~ E
    4.   (E V F) & (E V G)
    5.   F → (A & B)   2,   IMPL..
    6.   E → (~B & ~A)   3,   TRANS.
    7.   E → (~A & ~B)   6,   COM.
    8.   [F → (A & B)] & [E → (~A & ~B)]   5,7, CONJ.
    9   E V F   4   SIMP.
    10   F V E   9,   COM.
    11   (A & B) V (~A & ~B)   8,10, C.D.
    15. A↔ B   11, EQUIV.

5. 1.   [(~ A & B) & (E & F)] V (I & H)   ∴ A → I
    2.   (I & H) V [(~A & B) & (E & F)]   1,   COM.
    3.   [(I & H) V ~A & B)] &[(I & H) V(E & F)]2, DIST.
    4.   (I & H) V (~A & B)   3,   SIMP.
    5.   (H & I) V (~A & B)   4,   COM.
    6.   H & [I V (~A & B)   5,   ASSOC.
    7.   [I V (~A & B)] & H   6,   COM.
    8.   I V (~A & B)   7,   SIMP.
    9.   (I V ~A) & (I V B)   8,   DIST.
    10. I V ~A   9,   SIMP.
    11. ~A V I   10, COM.
    12. A → I   11, IMPL..

7. 1.   D → (C & F)   ∴ [(D → C) & (D → F)] & (A ↔ B)
    2.   B V ~ A
    3.   B → A
    4.   ~A V B   2,   COM.
    5.   A ↔ B   4,   IMPL..
    6.   (A → B) & (B → A)   5,3 CONJ.
    7.   A → B   6,   EQUIV.
    8.   ~D V (C & F)   1,   IMPL.
    9.   (~D V C) & (~D V F)   8,   DIST.
    10. (D → C) & (~D → F)   9,   IMPL.
    11. (D → C) & (D → F)   10, IMPL
    9.   [(D → C) & (D → F)] & (A↔ B)   12,7 CONJ.

DISTRIBUTION WORD PROBLEMS

3. 1.   E → (R & J)   ∴ E ↔ R
    2.   R → (E & B)
    3.   ~E V (R & J)   1,   IMPL..
    4.   (~E V R) & (~E V J)   3,   DIST.
    5.   ~E V R   4,   SIMP.
    6.   E → R   5,   IMPL..
    7.   ~R V (E & B)   2,   IMPL..
    8.   (~R V E) & (~R V E)   7,   DIST.
    9.   ~R V E   8,   SIMP.

10. R → E   9,   IMPL..
11. (E → R) & (R → E)   6,10, CONJ.
12. E↔R   11, EQUIV.

5. 1. (E & F) V [(N & B) V ~A]   ∴ ~E → N
    2. ~~A
    3. [(E & F) V (N & B)] V ~A   1,   ASSOC.
    4. ~A V [(E & F) V (N & B)]   3,   COM.
    5. (E & F) V (N & B)   4,2, D.S.
    6. [(E & F) V N] & [(E & F) V B]   5,   DIST.
    7. (E & F) V N   6,   SIMP.
    8. N V (E & F)   7,   COM.
    9. (N V E) & (N V F)   8,   DIST.
    10. N V E   9,   SIMP.
    11. E V N   10, COM.
    12. ~E → N   11, IMPL..

7. 1. M → (~B & P)   ∴ ~M
    2. B
    3. ~M V (~B & P)   1,   IMPL..
    4. (~M V ~B) & (~M V P)   3,   DIST.
    5. ~M V ~B   4,   SIMP.
    6. M → ~B   5,   IMPL..
    7. ~~B   2,   D.N.
    8. ~M   6,7, M.T.

CHAPTER 30: DE MORGAN'S THEOREMS

3. 1. P → Q   ∴ ~(P & ~Q)
    2. ~ P V Q   1,   IMPL..
    3. ~ (P & ~Q)   2,   DE M.

5. 1. F → ~B   ∴ ~ (A & B)
    2. ~E → F
    3. E → ~A
    4. E V F   2,   IMPL.
    5. (E → ~ A) & (F → ~ B)   3,1 CONJ.
    6. ~ A V ~ B   5,4 C.D.
    7. ~ (A & B)   6   DE M.

7. 1. A   ∴ ~[~(A V B) V ~(A V C)
    2. A V B   1   ADD.
    3. A V C   1   ADD.
    4. (A V B) & (A V C)   2,3 CONJ.
    5. ~ [~ (A V B) V ~ (A V C)]   5   DE M.

DE MORGAN WORD PROBLEMS

1. ~(~M & ~S)   ∴ ~M → S
    2. M V S   1,   DE M.
    3. ~M → S   2,   IMPL..

5. 1. ~(L & ~M)   ∴ L ↔ M
   2. ~(M & ~L)
   3. ~L V M   1,   DEM.
   4. ~M V L   2,   DE M.
   5. L → M   3,   IMPL..
   6. M → L   4,   IMPL..
   7. (L → M) & (M → L)   5,6   CONJ.
   8. L ↔ M   7, EQUIV.

7. 1. (P & A) → ~ (~C V ~D)   ∴ ~ D → ~ (P & A)
   2. (P & A) → (C & D)   1,   DEM.
   3. ~ (P & A) V (C & D)   2,   IMPL..
   4. [~(P & A) V C] & [~ (P & A) V D] 3,   DIST.
   5. ~ (P & A) V D   4,   COM; SIMP.
   8. ~D → ~(P & A)   5,   COM; IMPL.

CHAPTER 31: QUANTIFICATION THEORY
(page 243)

1. (X) (Lx → Cx)   2. (X) (Lx → ~ Cx)   3. (∃X) (Lx & Cx)
4. (∃X) (Lx & ~Cx) 5. (X) (Lx → Cx)   6. (X) (Cx → Lx)
7. (∃X) (Mx & Sx)   8. (∃X) (Mx & ~Sx)   9. (X) (Sx → Mx)
10. (X) (Sx → ~Mx)

QUANTIFICATION THEORY SYMBOLIC PROBLEMS

1. 1. (X) (Dx → Fx)   ∴ (X) (Dx → Cx)
   2. (X) (Fx → Cx)
   3. Da → Fa   1, UI.
   4. Fa → Ca   2, UI.
   5. Da → Ca   3,4 HS.
   6. (X) (Dx → Cx)   5, UG.

2. 1. (X) (Dx → Fx)   ∴ Fa
   2. Da
   3. Da → Fa   1, UI.
   4. Fa   3,2 MP.

3. 1. (X) (Dx → Fx)   ∴ (∃x) (Fx & Rx)
   2. (∃X) (Dx & Rx)
   3. Da → Ra   1, EI.
   4. Ra & Da   3, COM.
   5. Ra   4, SIMP.
   6. Da → Fa   1, UI.
   7. Da   3, SIMP
   8. Fa   6,7, M.P.
   9. Fa & Ra   8,5,   CONJ.
   10. (∃X)(Fx &Rx)   9, EG.

5. 1. (X) (Dx) → Fx)   ∴ (∃x) (Px & Rx)
   2   (X) (~Dx → Rx)
   3. (∃X) (~Fx & Px)

4.	~ Fa & Pa	3, EI.
5.	Da → Fa	1, UI.
6.	~ Fa	4, SIMP.
7.	~Da	5,6 MT.
8.	~ Da → Ra	2, UI.
9.	Ra	8,7 MP.
10.	Pa & ~ Fa	4, COM.
11.	Pa	10, SIMP.
12.	Pa & Ra	11.9 CONJ.
13.	(∃X) (Px & Rx)	12, EG.

7. 1. (X) (Ax → Bx)   ∴ (X) (~ EX → ~_Ax)
    2. (X) (~Cx → ~Bx)
    3. (X) (Cx → Dx)
    4. (X) (~Ex → ~Dx)
    5. ~Ea → ~Da   4 UI.
    6. ~Da → ~Ca   3 UI; TRANS.
    7. ~Ea → ~Ca   5,6 HS.
    8. ~Ca → ~Ba   2 UI.
    9. ~Ea → ~Ba   7,8 HS.
    10. ~Ba → ~Aa   1 UI.
    11. ~Ea → ~Aa   9,10 HS.
    12. (X) (~Ex → ~ Ax)   11 UG.

QUANTIFICATION THEORY WORD PROBLEMS

3. 1. (X) (Sx → Mx)   ∴ Sj → Mj
    2. Sj → Mj   1 UI.

5. 1. (X) (Sx → Px)   ∴ (X) (~ Px → ~ Sx)
    2. Sa → Pa   1 U.I.
    3. ~Pa → ~Sa   2 TRANS.
    7.   (X) (~ Px → ~ Sx) 3 UG.

7. 1. (X) (Ix → Mx)   ∴ ~ (Pj → ~ Mj)
    2. Pj & Ij
    3. Ij → Mj   1, UI.
    4. Ij   2, COM; SIMP.
    5. Mj   3,4 MP.
    6. Pj   2, SIMP.
    7. Pj & Mj   6,5, CONJ.
    8. ~ (~ Pj V ~ Mj)   7, DE M.
    9. ~ (Pj → ~ Mj)   8, IMPL.,

CHAPTER 32: FOUR BI-CONDITIONALS LOGICAL EQUIVALENCE

3. 1. ~(X) (Hx → ~Ix)   ∴ ~Ja
    2. (X) [Jx → (~Hx V ~Ix)
    3. (∃X) (Hx & Ix)   2, LE.
    4. Ha & Ia   3, EI.
    5. Ja → (~Ha V ~Ia) 2, UI.

6. Ja → ~(Ha & Ia)   5, DEM.
7. ~~(Ha & Ia)   4, D.N.
8. ~Ja   6,7   M.T.

5. 1. ~(∃X) (Hx & ~Fx)   ∴ ~ Ha
   2. ~(∃X) (Fx & Gx)
   3. Ga
   4. (X) (Hx → Fx)   1 LE.
   5. (X) (Fx →~Gx)   2 LE.
   6. Fa → ~Ga   5 UI.
   7. Ha → Fa   4 UI.
   8. ~~Ga   3 D.N.
   9. ~Fa   6,8 M.T.
   10. ~Ha   7,9 M.T.

6. 1. ~(X) (Bx → Cx)   ∴ ~ Sa
   2. ~(∃X) (Sx & ~Cx)
   3. (∃X) (Bx & ~Cx)   1 LE.
   4. (X) (Sx → Cx)   3 LE.
   5. Ba & ~Ca   3 EI.
   6. Sa → Ca   4 UI.
   7. ~Ca & Ba   5 COM.
   8. ~Ca   7 SIMP.
   9. ~Sa   6,8 M.T.

7. 1. (∃X) (Bx & Cx)   ∴ ~ (∃X) (Bx & Dx)
   2. ~(∃X) (~Cx & Dx)
   3. (X) (Bx → ~Cx)   1 LE.
   4. (X) (~Cx → ~Dx)   2 LE.
   5. Ba → ~Ca   3 UI.
   6. ~Ca → ~Da   4 UI.
   7. Ba → ~Da   5,6 H.S.
   8. (X) (Bx → ~Dx)   7 U.G.
   9. ~ (∃X) (Bx & Dx)   9 L.E.

FOUR BI-CONDITIONALS WORD PROBLEMS

3. 1. Aj   ∴ ~(X) (Ax → ~_Rx)
   2. Rj
   3. Aj & Rj   1,2 CONJ.
   4. (∃X) (Ax & Rx)   3, EG.
   5. ~(X) (Ax → ~Rx)

5. 1. ~ (∃X) (Ax & ~ Rx)   ∴ Rj
   2. (X) (Tx → Ax)
   3. Tj
   4. (X) (Ax →.Rx)   1, LE.
   5. Aj → Rj   4, UI.
   6. Tj → Aj   2, UI.
   7. Tj → Rj   6,5 HS.
   8. Rj   7,3 MPL.

7. 1. ~ (∃X) (Sx & ~_Mx)   ∴ ~ (∃X) (Sx & Zx)
   2. ~ (∃X) (Mx & ~ Tx)
   3. (X) (Zx → Px)
   4. ~ (∃X) (Px & Tx)
   5. (X) (Sx → MX)   1   LE.
   6. (X) (Mx → Tx)   2   LE.
   7. (X) (Px → ~Tx)   4   LE.
   8. Sa → Ma   5   UI.
   9. Ma → Ta   6   UI.
   10. Pa → ~Ta   7   UI.
   11. Za → Pa   3   UI.
   12. Ta → ~Pa   10 TRANS.
   13. ~Pa → ~_Za   11 TRANS.
   14. Sa → Ta   8,9 HS.
   15. Sa → ~_Pa   14,12 HS.
   16. Sa → ~_Za   15,13 HS.
   17. (X) (Sx → ~_Zx)   16 UG
   18. ~ (∃X) (Sx & Zx)   17 LE.